Systems & Control: Foundations & Applications

Series Editor
Tamer Başar, University of Illinois at Urbana-Champaign, Urbana, IL, USA

Editorial Board
Karl Johan Åström, Lund University of Technology, Lund, Sweden
Han-Fu Chen, Academia Sinica, Beijing, China
Bill Helton, University of California, San Diego, CA, USA
Alberto Isidori, Sapienza University of Rome, Rome, Italy
Miroslav Krstic, University of California, San Diego, CA, USA
H. Vincent Poor, Princeton University, Princeton, NJ, USA
Mete Soner, ETH Zürich, Zürich, Switzerland;
    Swiss Finance Institute, Zürich, Switzerland
Roberto Tempo, CNR-IEIIT, Politecnico di Torino, Italy

For further volumes:
http://www.springer.com/series/4895

Leonid Fridman • Alexander Poznyak
Francisco Javier Bejarano

Robust Output LQ Optimal Control via Integral Sliding Modes

Leonid Fridman
Departamento de Ingeniería
  de Control y Robótica
Universidad Nacional Autonoma
  de Mexico
Mexico City, Distrito Federal, Mexico

Alexander Poznyak
Department of Control Automatico
Centro de Investigacion y Estudios
  Avanzados (CINVESTAV)
Mexico City, Distrito Federal, Mexico

Francisco Javier Bejarano
Department of Research
  and Posgraduates Studies (SEPI)
ESIME Ticomán, Instituto
  Politécnico Nacional
Mexico City, Distrito Federal, Mexico

ISSN 2324-9749   ISSN 2324-9757 (electronic)
ISBN 978-0-8176-4961-6   ISBN 978-0-8176-4962-3 (eBook)
DOI 10.1007/978-0-8176-4962-3
Springer New York Heidelberg Dordrecht London

Library of Congress Control Number: 2014930565

Mathematics Subject Classification (2010): 93C73, 93B12, 49Kxx, 90C47, 49K35, 93B07

© Springer Science+Business Media New York 2014
This work is subject to copyright. All rights are reserved by the Publisher, whether the whole or part of the material is concerned, specifically the rights of translation, reprinting, reuse of illustrations, recitation, broadcasting, reproduction on microfilms or in any other physical way, and transmission or information storage and retrieval, electronic adaptation, computer software, or by similar or dissimilar methodology now known or hereafter developed. Exempted from this legal reservation are brief excerpts in connection with reviews or scholarly analysis or material supplied specifically for the purpose of being entered and executed on a computer system, for exclusive use by the purchaser of the work. Duplication of this publication or parts thereof is permitted only under the provisions of the Copyright Law of the Publisher's location, in its current version, and permission for use must always be obtained from Springer. Permissions for use may be obtained through RightsLink at the Copyright Clearance Center. Violations are liable to prosecution under the respective Copyright Law.
The use of general descriptive names, registered names, trademarks, service marks, etc. in this publication does not imply, even in the absence of a specific statement, that such names are exempt from the relevant protective laws and regulations and therefore free for general use.
While the advice and information in this book are believed to be true and accurate at the date of publication, neither the authors nor the editors nor the publisher can accept any legal responsibility for any errors or omissions that may be made. The publisher makes no warranty, express or implied, with respect to the material contained herein.

Printed on acid-free paper

Springer is part of Springer Science+Business Media (www.birkhauser-science.com)

*We dedicate this book with love and gratitude to
Leonid's wife Millie,
Alexander's wife Tatyana,
and
Francisco's mother Maria Acela*

Preface

The idea of this research arose from the discussion between Professors Y. Shtessel, A. Poznyak, and L. Fridman in Mexico City in September 2002 motivated by the recently obtained result, the so-called robust min–max principle, by Professors V. Boltyanski and A. Poznyak. The main topic of this discussion was about the advantages and disadvantages of two different types of robust control concepts: min–max and sliding modes. In conclusion, Professors Poznyak and Fridman decided to start the investigation joining the advantages of two approaches to robustness:

- the ability of integral sliding mode controllers to compensate matched uncertainties starting from initial time moment;
- the possibilities of robust optimal controllers to provide the best possible solution for the worst case of a set of uncertainties.

This decision finally defined the topic of master and Ph.D. thesis and postdoctoral study life of Dr. Francisco Javier Bejarano.

Working on this topic we discover that for the case when the number of matched unknown inputs is less than the number of the outputs sometimes it is possible to design an observer estimating unmeasured coordinates theoretically exactly starting from the initial time moment even in the presence of unknown inputs.

In this way the concept of *output integral sliding modes* was born which we would like to present in this book.

Mexico City, Mexico

Leonid Fridman
Alexander Poznyak
Francisco Javier Bejarano

Acknowledgements

We wish to thank our friends and colleagues Professors Vadim Utkin, Yury Shtessel, and Michael Basin for stimulating discussions benefiting the book.

We also thank Dr. Alejandra Ferreira, Dr. Manuel Jimenez, Dr. Liset Fraguela, and Alfredo Sosa for simulations, supporting experiments, and editorial work.

The book was partially supported by the grants of Consejo Nacional de Ciencia y Tecnología de México 132125.

Contents

1 **Introduction** 1
   1.1 Importance of Robust Control 1
   1.2 Min–Max Concept 2
   1.3 Sliding Mode Control 2
      1.3.1 Main Steps of Sliding Mode Control 2
      1.3.2 Sliding Mode Control for Systems with Unmatched Uncertainties/Disturbances 3
      1.3.3 Output-Based Minimization of Unmatched Uncertainties/Disturbances via SMC 3
   1.4 Integral Sliding Mode Control 4
      1.4.1 Main Results 4
   1.5 Main Contribution of the Book 5
   1.6 Structure of the Book 5
   1.7 How to Read This Book? 7

Part I OPTIMAL CONTROL AND SLIDING MODE

2 **Integral Sliding Mode Control** 11
   2.1 Motivation 11
   2.2 Problem Formulation 12
   2.3 Control Design Objective 13
   2.4 ISM Control Design 13
   2.5 Linear Case 14
   2.6 Example: LQ Optimal Control and ISM 15
      2.6.1 Unmatched Disturbances 16
      2.6.2 Example: ISM and Unmatched Disturbances 19
   2.7 Conclusions 19

x   Contents

3   Observer Based on ISM 21
   3.1   Motivation 21
   3.2   System Description 22
   3.3   Observer Design.. 22
        3.3.1   Auxiliary Dynamic Systems and Output Injections 23
   3.4   Observer in the Algebraic Form 26
   3.5   Observer Realization 28
   3.6   Example 28

4   Output Integral Sliding Mode Based Control 31
   4.1   Motivation 31
   4.2   System Description 32
   4.3   OISM Control 32
   4.4   Output Integral Sliding Modes 33
   4.5   Design of the Observer 34
   4.6   LQ Control Law.. 36
   4.7   Example 37
   4.A   Proof of Lemma 4.1.. 38
   4.B   Proof of Lemma 4.2.. 41

Part II MIN–MAX OUTPUT ROBUST LQ CONTROL

5   The Robust Maximum Principle 45
   5.1   Min–Max Control Problem in the Bolza Form 45
        5.1.1   System Description 45
        5.1.2   Feasible and Admissible Control..................... 46
        5.1.3   The Cost Function and the Min–Max
              Control Problem 47
        5.1.4   The Mayer Form Representation 48
        5.1.5   The Hamiltonian Form 48
   5.2   Robust Maximum Principle 50
        5.2.1   Main Result 50
   5.3   Min–Max Linear Quadratic Multimodel Control 51
        5.3.1   The Problem Formulation 51
        5.3.2   The Hamiltonian Form and the Parameterization
              of Robust Optimal Controllers 51
        5.3.3   The Extended Form for the Closed-Loop System 53
        5.3.4   The Robust LQ Optimal Control 54
        5.3.5   Robust Optimal Control for Linear Stationary
              Systems with Infinite Horizon 56
   5.4   Conclusions.. 57

6 Multimodel and ISM Control 59
- 6.1 Motivation 59
- 6.2 Problem Formulation 60
- 6.3 Design Principles 61
- 6.4 Optimal Control Design 63
- 6.5 Examples 64
- 6.6 Linear Time Invariant Case 66
  - 6.6.1 Transformation of the State 68
  - 6.6.2 The Corrected LQ Index 68
  - 6.6.3 Min–Max Multimodel Control Design 69
- 6.7 Example 73

7 Multiplant and ISM Output Control 77
- 7.1 Motivation 77
- 7.2 Problem Formulation 78
- 7.3 Output Integral Sliding Mode 79
- 7.4 Design of the Observer 81
  - 7.4.1 Auxiliary Dynamic Systems and Output Injections 82
  - 7.4.2 Observer in Its Algebraic Form 84
  - 7.4.3 Observer Realization 85
- 7.5 Min–Max Optimal Control Design 86
  - 7.5.1 Control Algorithm 87
- 7.6 Error Estimation During Implementation of the Closed-Loop Control 88
- 7.7 Example 89

Part III PRACTICAL EXAMPLES

8 Fault Detection 97
- 8.1 Model Description 97
- 8.2 Observer Design 98
- 8.3 Fault Estimation 100

9 Stewart Platform 103
- 9.1 Model Description 103
- 9.2 Output Integral Sliding Mode 106
- 9.3 Min–Max Stabilization of Platform P 107
- 9.4 Numerical Simulations 109

10 Magnetic Bearing 115
- 10.1 Preliminaries 115
- 10.2 Disturbances Compensator 117
- 10.3 Observer Design 118
- 10.4 Numerical Simulations 119

A	**Sliding Modes and Equivalent Control Concept** 123	
	A.1 Introduction 123	
	A.2 Equivalent Control Method 124	
B	**Min–Max Multimodel LQ Control** 131	
	B.1 Multimodel System 131	
	B.2 Numerical Method for the Weight Adjustment 133	
	B.2.1 Numerical Method................................. 136	
	B.3 Example 138	

Notations 143

References 145

Index 149

1
Introduction

1.1 Importance of Robust Control

Robust control is a branch of modern control theory that explicitly deals with uncertainty in its approach to controller design. Robust control methods are designed to function properly so long as uncertain parameters or disturbances are within some (typically compact) set. Robust methods aim to achieve robust performance and/or stability in the presence of bounded modelling errors. The classical control design, based on the frequency domain methodology, was fairly robust; the state-space methods invented in the 1960s and 1970s were sometimes found to lack robustness [1], prompting research to improve them. This was the start of the theory of robust control, which took shape in the 1980s and 1990s and is still active today. In contrast with an adaptive control policy, a robust control policy is static; rather than adapting to measurements of variations, the controller is designed to work assuming that certain variables will be unknown but, for example, bounded [2, 3].

When is a control method said to be robust? Informally, a controller designed for a particular set of parameters is said to be robust if it would also work well under a different set of assumptions. High-gain feedback is a simple example of a robust control method; with sufficiently high gain, the effect of any parameter variations will be negligible. High-gain feedback is the principle that allows simplified models of operational amplifiers and emitter-degenerated bipolar transistors to be used in a variety of different settings. This idea was already well understood by Bode and Black in 1927.

The modern theory of robust control began in the late 1970s and early 1980s and soon developed a number of techniques for dealing with bounded system uncertainty [4, 5]. Probably the most important example of a robust control technique is H^∞ loop-shaping, which was developed by Duncan McFarlane and Keith Glover [6]; this method minimizes the sensitivity of a system over its frequency spectrum, and this guarantees that the system will have sufficiently small deviation from expected trajectories when disturbances enter the system. Another example is LQG/LTR, which was developed to overcome the robustness problems of LQG control [7]. In [8] the polynomial robust stability is analyzed.

In this book we will deal with two different types of robust control strategies: *min–max concept* and *sliding mode control*. Let us discuss firstly the advantages and drawbacks of both strategies.

1.2 Min–Max Concept

When we do not have a complete information on a dynamic model to be controlled the main problem consists in designing an acceptable control which remains to be "close to an optimal one" having a small sensibility with respect to any unknown (unpredictable) factor from a given possible set. In other words the desired control should be *robust* with respect to an unknown factor. In the presence of any sort of uncertainties (parametric type, unmodelled dynamics, external perturbations, etc.) the main way to obtain a solution suitable for a class of given models is to formulate a corresponding *min–max control* problem, where

- *maximization* is taken over a set of uncertainty;
- *minimization* is taken over control actions within a given set.

The min–max controller design for different classes of nonlinear systems has been a hot topic of research for over the last two decades. A recent more comprehensive publication on this topic can be found in [9].

Three drawbacks of min–max concept are that:

- it requires the availability of the entire state vector along all the time;
- it deals only with parametric uncertainties;
- it needs the complete knowledge of all possible plant variations.

1.3 Sliding Mode Control

1.3.1 Main Steps of Sliding Mode Control

Sliding mode controllers (SMC) were developed in the Soviet Union in the mid-1950s (see, e.g., [10]) in the framework of variable structure control (VSC), a nonlinear control method that alters the dynamics of a nonlinear system by the application of a switching control. In the framework of VSC it was understood that if the controllers are ensuring finite-time arrival to some surface in both sides of the surface, the solution should slide on the surface if it is supposed the frequency of the switching is infinite. Moreover, analyzing the phase plane, it was shown that such motions have three principal specific features (see, e.g., [11–14]):

- the sliding mode dynamics are not coinciding with any dynamics of the system outside of the surface;
- the sliding motions are invariant with respect to uncertainties/disturbances;
- the sliding dynamics is described by reduced order equations;
- the finite-time convergence of the system trajectory to the sliding surface.

Later in the early 1960s the rigorous mathematical analysis of SMC was done (see [15]). In 1969 Drazenovic [16] showed that the sufficient and necessary condition that sliding dynamics be invariant with respect to uncertainties/disturbances be that they should be matched. In 1981 Lukyanov and Utkin [17] proposed a two-step procedure of SMC design:

- design of the sliding surface;
- discontinuous controller design.

However, SMC has the following drawbacks:

- chattering, i.e., fast undesirable oscillations inspired by discontinuity of control law and presence of nonidealities: parasitic unmodelled dynamics, hysteresis and time delays, etc.;
- sliding motions are invariant with respect to matched perturbations only.

1.3.2 Sliding Mode Control for Systems with Unmatched Uncertainties/Disturbances

The SMC for the systems with unmatched uncertainties are designed in many papers.

We would like to underline the following direction:

1. Compensation of unmatched uncertainties/disturbances using dynamic sliding surfaces is presented in [18] (see also a discussion therein).
2. The LMI-based approach is applied in [19].
3. The combination of backstepping and higher-order sliding modes [20, 21].
4. [22] proposed the LQ multimodel problem solution presented as a combination of two optimal problems: firstly an optimal sliding surface for singular Multimodel LQ problem was designed. After that, the time minimization problem for reaching phase was solved.

The main disadvantage of those approaches is that they need complete information about system states.

1.3.3 Output-Based Minimization of Unmatched Uncertainties/Disturbances via SMC

Normally the output-based sliding mode controllers are designed basing on some kind of observers. In doing so:

- output-based minimization of unmatched uncertainties/disturbances using H^∞ was proposed in [23, 24];
- in [25] the observer-based approach was suggested identifying the perturbations and compensating them through sliding surface;

In all the abovementioned approaches sliding motions are not starting from initial time moment, i.e., the reaching phase exists and does not allow the matched uncertainties/disturbances compensation from initial time moment.

1.4 Integral Sliding Mode Control

1.4.1 Main Results

In some control problems the control law, i.e., the nominal trajectory, is already done in the initial state space. The only thing the designers need is to ensure the insensitivity of the trajectory tracking with respect to uncertainties starting from the initial time moment. To ensure exact (with respect to the matched uncertainties/disturbances) tracking of the nominal trajectory designed for nominal systems in original state space starting from initial time moment the concept of *integral sliding mode control* (ISMC) [26, 27] was proposed.

The integral sliding surface is a surface in an extended state space. The motions on this surface are starting from the initial time moment. So the systems governed by ISMC have the following advantages:

- compensation of the matched uncertainties/disturbances starts from the initial time moment since the motion surface is a virtual surface;
- the motions in integral sliding modes have the same dimension as the initial state space;
- it leads to chattering reduction, because ISMC needs the smaller discontinuous control gains since the nominal system dynamics are supposed to be already compensated by the nominal control law.

Unfortunately the main drawbacks of ISMC are that:

- they need complete information about all of the system's states starting from initial time moment;
- ISMC cannot compensate unmatched uncertainties.

1.4.1.1 ISM-Based Compensation of Unmatched Uncertainties

The works [28, 29] presented a projection method allowing to design ISMC compensating completely matched uncertainties/disturbances and minimizing and not amplifying the unmatched once.

In the papers [28, 30–32] the combination of ISMC and H^∞ control is suggested for minimization of the effect of the presence of unmatched uncertainties/disturbances on the quality of nominal trajectory tracking.

1.5 Main Contribution of the Book

The aim of the book is to present a concept of *output integral sliding modes* (OISM). OISM [33, 34] controller provides theoretically exact tracking of nominal trajectory for the systems with matched uncertainties if we suppose that the ideal sliding modes do exist and equivalent control signals are available.

This concept has two main advantages:

- it provides the information about the system states:
  - theoretically exactly;
  - right after initial time moment;
  - even in the presence of matched uncertainties;
- it ensures exact tracking of the nominal trajectory:
  - right after the first moment;
  - in the presence of matched uncertainties;
  - basing on output's information only.

Combination of OISM and LQ controllers allows maybe firstly in the history to offer theoretically exact solution of LQ problem basing on output's information only.

Application of OISM to robustification of LQ problem for linear uncertain systems ensures theoretically exact tracking of nominal LQ trajectory:

- in the presence of matched uncertainties;
- starting from initial time moment;
- using only output information.

Combination of OISM with multimodel LQ problem has one more advantage: it allows to eliminate the matched part of model variations and uncertainties and consider only the unmatched part.

In Table 1.1, we have summarized the main advantages of the combinations of ISM and OISM strategies with LQ and multimodel LQ problems for LTV uncertain systems. The advantages of the proposed strategies are marked in bold letters.

1.6 Structure of the Book

The book consists of an introduction, three parts, and two appendixes.

In Part I the concept of *output integral sliding modes* is presented.

As the first step in Chap. 2 the concept of integral sliding mode (ISM) is revisited. The efficiency of ISM is illustrated on the example of robustification of LQ control for systems with uncertainties/disturbances. Then the projection to the unmatched variable subspace is designed ensuring that by applying the ISMC we are not amplifying, but minimizing the unmatched uncertainties/disturbances.

Table 1.1. Advantages of using min–max and OISM together

	Unmatched model variation	Matched uncertainties /disturbances	Needed information
MM-LQ	Part of the min–max problem in original space	Can't reject	All states
ISM+LQ	Can't reject	**Compensates completely**	All states
OISM+LQ	Can't reject	**Compensates completely**	**Output**
MM-LQ + OISM	**Eliminates matched variations of models**	**Compensates completely**	**Output**

Chapter 3 describes the OISM observer design. The main idea of such observer is the following: by designing the output-based integral sliding mode one can reconstruct the value of the output´s derivative as an equivalent control signal right after initial time moment, provided that ideal sliding modes do exist and the equivalent control value is available. Applying such procedure step-by-step one can reconstruct all necessary derivatives of the outputs and consequently observe theoretically exactly all of the system states for observable systems right after initial time moment.

Chapter 4 presents the main concept of the book *output integral sliding mode control*. Such type of controllers ensures theoretically *exact tracking* of nominal optimal trajectory *right after the initial moment* even in the presence of matched uncertainties/disturbances *based on output information only*, if it is supposed that there exist ideal sliding modes and equivalent output signal is available. The discrete realization of output integral sliding mode controller requires the filtration to obtain the equivalent output injections. It is shown that the observation error can be made arbitrarily small after an arbitrary small time without any adjustment of the observer parameters, only by decreasing the sampling step and filter time constant.

Part II presents three different combinations of min–max control [9] and ISMC or OISM control.

This part starts with Chap. 5 revisiting the concept of robust min–max control. Firstly the min–max control problem in Bolza form is discussed. Then the robust exact principle theorem is formulated. Finally, the application of robust maximum principle to the solution of LQ problem for multimodel systems is given.

Chapter 6 presents the combination of ISMC with min–max controllers basing on the state information. The multimodel systems with matched uncertainties are considered. It is shown that the application of ISMC to the solution of min–max problem reduces to a solution of an equivalent min–max nominal LQ problem. The ISMC completely dismisses the influence of matched uncertainties right after the initial time instant.

In Chap. 7 we consider the application of a min–max optimal control based on the LQ index for a set of systems where only the output information is available. Here every system is affected by matched uncertainties, and we propose to use an OISM to compensate matched uncertainties right after the beginning of the process if we suppose that there exist ideal sliding modes and equivalent output injections are available. For the case when the extended system is free of invariant zeros, a hierarchical sliding mode observer is applied. The error of realization of the proposed control algorithm is estimated in terms of the sampling step and actuator time constant.

Part III of the book presents applications of the methodology developed in Part II to three different control and observation problems.

In Chap. 8 an OISM-based fault detection scheme is proposed. The efficiency of the proposed scheme is illustrated by the example of the estimation of the actuator's level damage in the cart pendulum.

Chapter 9 tackles the problem of a two-player differential game affected by matched uncertainties with only the output measurement available for each player. We suggest a state estimation based on the so-called algebraic hierarchical observer for each player in order to design the Nash equilibrium strategies based on such estimation. At the same time, the use of an output integral sliding mode term for the Nash strategies robustification for both players ensures the compensation of the matched uncertainties. A simulation example shows the feasibility of this approach in a magnetic levitator problem.

In Chap. 10 the OISM controllers, based only on output information, are applied to a Stewart platform. This platform has three degrees of freedom, and it is used as a remote surveillance device. We consider the hierarchical sliding mode observer, allowing the reconstruction of the system states from the initial moment. This allows the implementation of an OISM controller ensuring the insensitivity of the state trajectory with respect to the matched uncertainties from the initial moment.

In Appendixes we present the most important material needed to read the book. Appendix A presents basic information about equivalent control method for definition of solution in sliding mode. There, a lemma, by [11], about online calculation of the equivalent control is presented. Appendix B presents a numerical method for the optimal weight adjustment for the min–max LQ problem, where "max" is taken over a finite set of indices (models) and "min" is taken over the set of admissible controls. The solution is obtained by the robust optimal control application. The control turns out to be a linear combination of the controls optimal for each individual model.

We hope that such structure makes the book complete and self-content.

1.7 How to Read This Book?

In writing the book we supposed that it can be useful for readers interested in:

- robustification of optimal control problems;
- new methods of sliding mode control.

So we supposed that we will have four different categories of readers:

- beginners;
- readers skilled in optimal control;
- readers skilled in sliding mode control;
- readers skilled in both optimal and sliding mode control.

So we would like to suggest four strategies in reading this book:

1. It is desirable that beginners have the basic knowledge about
   - LQ control (e.g., [35]);
   - sliding mode control (see Appendix A containing the minimal necessary information or for more deep knowledge [11, 12, 14]).

   After these two steps this category of readers can start with Chap. 1.
2. Readers skilled in optimal control should read firstly Appendix A containing the minimal necessary information or for more deep knowledge [11, 12, 14] and then start to read the book.
3. Readers skilled in sliding mode control should revise the basic books of LQ control (e.g., [35]).
4. Readers skilled in both optimal and sliding mode control can start to read the book from Part I.

For the readers which would like to use the book results for implementation we have included Appendix B discussing numerical realization for min–max multimodel control.

Enjoy reading!

Part I

OPTIMAL CONTROL AND SLIDING MODE

2
Integral Sliding Mode Control

Abstract In this chapter the concept of integral sliding mode (ISM) is revisited. The efficiency of ISM is illustrated on the example of LQ control. Then the projection to the unmatched variable subspace is designed ensuring that the application of ISM is not amplifying, but minimizing the unmatched perturbations. An illustrative example of application of ISM to LQ problem is presented.

2.1 Motivation

Sliding mode control techniques are very useful for the controller design in systems with disturbances and model/parametric uncertainties. The system's compensated dynamics become insensitive to matched disturbances and uncertainties under sliding mode control. The price for this insensitivity is control chattering and a reaching phase, during which the system's dynamics are vulnerable to disturbances/uncertainties. For linear systems, whose dynamics are completely known, a traditional controller, including proportional-plus-derivative (PD), proportional-plus-integral-plus-derivative (PID), and optimal linear quadratic regulator (LQR), can be successfully designed to compensate the dynamics. A nonlinear system which is completely known can be compensated, for instance, by a feedback linearization controller, backstepping controller, or any other Lyapunov-based nonlinear controller [36]. Systems compensated by these types of controllers will be of the full order equal to the order of the uncompensated system. Once the system is subjected to external bounded disturbances, it is natural to try to compensate such disturbances by means of an auxiliary control retaining the effect of the main controller designed for the unperturbed system. The sliding mode based auxiliary controller that compensates the disturbance from the very beginning of the control action, while retaining the order of uncompensated system, is named integral sliding mode (ISM) controller. This chapter is dedicated to the study of the ISM controller design. ISM has been studied in [13, 26–28, 33, 37–40].

2.2 Problem Formulation

Consider the following controlled uncertain system represented by the state-space equation:

$$\dot{x}(t) = f(x(t)) + B(x(t))u(t) + \phi(x,t) \tag{2.1}$$

where $x(t) \in \mathbb{R}^n$ is the state vector and $u(t) \in \mathbb{R}^m$ is the control input vector. The function $\phi(x,t)$ represents the uncertainties affecting the system due to parameter variations, unmodelled dynamics, and/or exogenous disturbances.

Let $u = u_0$ be a nominal control designed for (2.1) assuming $\phi = 0$, where u is designed to achieve a desired task, whether it be stabilization, tracking, or an optimal control problem. Thus, the trajectories of the ideal system ($\phi = 0$) will be given by the solutions of the following ODE equations:

$$\dot{x}_0(t) = f(x_0(t)) + B(x_0(t))u_0(t) \tag{2.2}$$

Thus, for $x(0) = x_0(0)$ and ϕ being not equal to zero, the trajectories of (2.1) and (2.2) are different. The trajectories of (2.2) satisfy some specified requirements, whereas the trajectories of (2.2) might have a quite different performance (depending on ϕ) to the one expected by the control designer. For the control design given below it is necessary to assume that:

A3.1. rank $B(x) = m$ for all $x \in \mathbb{R}^n$;
A3.2. the disturbance $\phi(x,t)$ is assumed to be matched, i.e., it satisfies the so-called *matching condition*:

$$\phi(x,t) \in \operatorname{Im} B(x)$$

i.e., there exists a vector $\gamma(x,t) \in \mathbb{R}^m$ such that $\phi(x,t) = B(x)\gamma(x,t)$.

- From a control point of view, the matching condition means that the effects produced by $\phi(x,t)$ in the system can be produced by u, and vice versa.

A3.3. An upper bound for $\gamma(x,t)$ can be found, i.e.,

$$\|\gamma(x,t)\| \leq \gamma^+(x,t) \tag{2.3}$$

Obviously, the second restriction is needed to compensate ϕ; if it is known, it would be enough to choose $u = -\gamma$. However, since γ is uncertain, some other restrictions are needed in order to eliminate the influence of ϕ. In this way, the sliding mode approach replaces the lack of knowledge of ϕ by the first and third assumptions.

2.3 Control Design Objective

Now the control design problem is *to design a control law* that, provided that $x(0) = x_0(0)$, *guarantees the identity* $x(t) = x_0(t)$ *for all* $t \geq 0$. By comparing (2.1) and (2.2), it is clear that the control design is achieved only if the equivalent control is equal to the negative of the uncertainty ($u_{1_{eq}} = -\gamma$). Thus, the control objective can be reformulated in the following terms: *design the control* $u = u(t)$ *in the following form:*

$$u(t) = u_0(t) + u_1(t) \quad (2.4)$$

where $u_0(t)$ is the nominal control part designed for (2.2) and $u_1(t)$ is the *integral sliding mode* (ISM) control part guarantying the compensation of the unmeasured matched uncertainty $\phi(x,t)$, starting from the beginning ($t=0$).

2.4 ISM Control Design

Since $\phi(x,t) = B(x)\gamma(x,t)$, substitution of (2.4) into (2.1) yields

$$\dot{x} = f(x) + B(x)(u_0 + u_1 + \gamma)$$

The sliding manifold is given by means of the equation $s(x) = 0$ with s defined by the formula

$$s(x) = s_0(x) - s_0(x(0)) - \int_0^t G(x(\tau))\left[f(x(\tau)) + B(x(\tau))u_0(\tau)\right]d\tau \quad (2.5)$$

where $s_0(x) \in \mathbb{R}^m$ is a vector that could be designed as a linear combination of the state and $G(x) = \frac{\partial s_0}{\partial x}$. Then, in contrast with conventional sliding modes, here an integral term is included. Furthermore, in this case we have $s(x(0)) = 0$.

Thus, the time derivative of s is obtained by the formula

$$\dot{s} = G(x)B(x)(u_1 + \gamma)$$

In order to achieve the sliding mode, the term s_0 should be designed such that

$$\det[G(x)B(x)] \neq 0, \text{ for all } x \in \mathbb{R}^n$$

The sliding mode control should be designed as

$$u_1 = -M(x,t)\frac{D^T s}{\|D^T s\|} \quad (2.6)$$

$$M(x,t) > \gamma^+(x,t), \quad D(x) = G(x)B(x)$$

Taking $V = \frac{1}{2}s^T s$, and in view of (2.3), the time derivative of V is bounded as follows:

$$\dot{V} = (s, \dot{s}) = (s, D(u_1 + \gamma)) = (D^T s, u_1 + \gamma)$$
$$\leq -\|D^T s\| (M - \gamma^+) < 0$$

Hence V decreases, which implies

$$V(t) \leq V(0) = \frac{1}{2}\|s(x(0))\|^2 = 0$$

That is, **the sliding mode is achieved from the beginning.** Now, the equivalent control $u_{1_{eq}}$ is taken from $\dot{s} = 0$

$$\dot{s} = u_1 + \gamma = 0$$

A review of the equivalent control method is discussed in Appendix A. As it is explained there, $u_{1_{eq}}$ is taken as the solution for the control obtained from the equation of $\dot{s}$ when this is equal to zero. Thus, in this case,

$$u_{1_{eq}} = -\gamma$$

Hence, the sliding motion is given by

$$\dot{x}(t) = f(x(t)) + B(x(t)) u_0(t)$$

and our aim is achieved since now $x(t) \equiv x_0(t)$.
Notice that the order of the dynamic equation in the sliding mode is not reduced. This property defines an *integral sliding mode* [13].

2.5 Linear Case

Let us consider the linear time invariant system:

$$\dot{x} = Ax + B(u_0 + u_1) + \phi \qquad (2.7)$$

In this case the vector function s can be defined by means of the formula

$$s(x) = G(x(t) - x(0)) + (GB)^{-1} G \int_0^t (Ax(\tau) + Bu_0(\tau)) d\tau \qquad (2.8)$$

where $G \in \mathbb{R}^{m \times n}$ is a projection matrix satisfying the condition

$$\det[GB] \neq 0$$

Thus, the time derivative of s takes the form

$$\dot{s}(x) = GB(u_1 + \gamma)$$

The control u_1 is designed as

$$u_1 = -M(x,t) \frac{(GB)^T s}{\left\|(GB)^T s\right\|} \qquad (2.9)$$

$$M(x,t) > \gamma^+(x,t)$$

Therefore, taking $V = \frac{1}{2}s^T s$, and in view of (2.3), the following inequality is obtained:

$$\dot{V} = \left((GB)^T s, \dot{s}\right) = (s, u_1 + \gamma)$$
$$\leq -\left\|(GB)^T s\right\|(M - \gamma^+) < 0$$

Hence, the integral sliding mode is guaranteed.

2.6 Example: LQ Optimal Control and ISM

Consider the following system:

$$\dot{x} = Ax + B(u_0 + u_1) + \phi$$

representing a linearized model of an inverted cart–pendulum of Fig. 2.1, where x_1 and x_2 are the car position and pendulum angle and x_3 and x_4 are their respective velocities. The matrices A and B take the following values:

$$A = \begin{bmatrix} 0 & 0 & 1 & 0 \\ 0 & 0 & 0 & 1 \\ 0 & 1.25 & 0 & 0 \\ 0 & 7.55 & 0 & 0 \end{bmatrix}, \quad B = \begin{bmatrix} 0 \\ 0 \\ 0.19 \\ 0.14 \end{bmatrix}$$

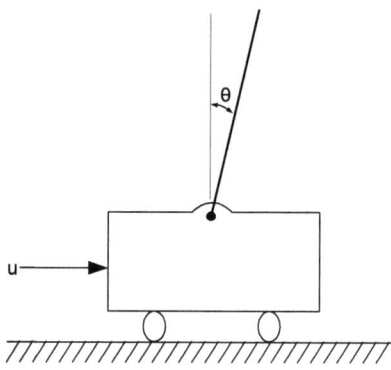

Fig. 2.1. Inverted cart–pendulum.

The control $u_0 = u^*$ is designed for the nominal system, where u^* solves the following optimal problem subject to an LQ performance index:

$$J(u_0) = \int_0^\infty x_0^T(t) Q x_0(t) + u_0(t)^T R u_0(t) \, dt$$

$$u_0^* = \arg\min J(u_0)$$

It is known (see, e.g., [35]) that the solution of the previous optimal control is given in its state feedback representation by means of

$$u_0^*(x) = -R^{-1} B^T P x$$

where P is a symmetric positive definite matrix that is the solution of the algebraic Riccati equation

$$A^T P + PA - PBR^{-1} B^T P = -Q$$

For the considered matrices A and B, and taking $Q = I$ and $R = 1$, we have that P and $K := R^{-1} B^T P$ have the following values:

$$P = \begin{bmatrix} 4.3 & -48.5 & 8.9 & -18.9 \\ -48.5 & 3149 & -191.4 & 1174.4 \\ 8.9 & -191.4 & 33.1 & -74.5 \\ -18.9 & 1174.4 & -74.5 & 438.6 \end{bmatrix}$$

$$K = \begin{bmatrix} -1 & 131.36 & -4.337 & 48.47 \end{bmatrix}$$

We considered that $\phi = B\gamma$ with $\gamma = 2\sin(0.5t) + 0.1\cos(10t)$ and the ISM control is

$$u_1 = -5 \operatorname{sign}(s)$$

where s is designed according to (2.8). Now, the only restriction over G is that $\det GB \neq 0$ and therefore we have a big range of election. One simple selection is $G = \begin{bmatrix} 0 & 0 & 1 & 0 \end{bmatrix}$, and thus we obtain $GB = 0.19$, which obviously is different from zero.

Figure 2.2 shows the position of the car and the pendulum. We can see that there is no influence of the disturbance γ thanks to the compensation effect caused by the ISM control part u_1.

2.6.1 Unmatched Disturbances

One may think why not use $G = B^+ = (B^T B)^{-1} B^T$. In such a way $\dot{s} = (u_1 + \gamma)$ and control u_1 is still as in (2.9). A criterion for selecting G in an appropriate way can be given if we do not assume $\phi(x,t)$ to be matched (it may or may not be).

Let $B^\perp \in \mathbb{R}^{n \times m}$ be a full rank matrix whose image is orthogonal to the image of B, i.e., $B^T B^\perp = 0$ and $\begin{bmatrix} B & B^\perp \end{bmatrix}$ is nonsingular. Notice that

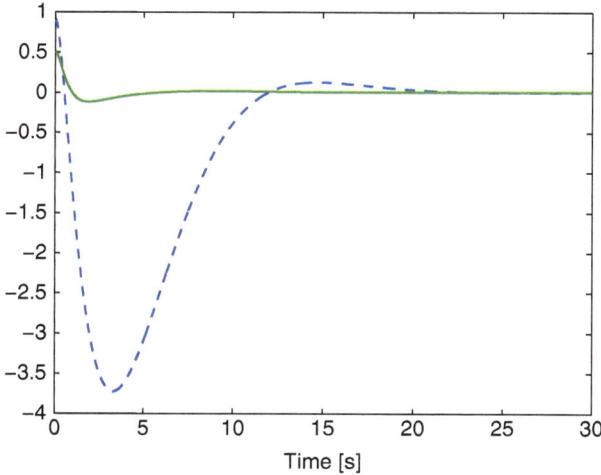

Fig. 2.2. States x_1 (*dashed*) and x_2 (*solid*) using ISM for matched uncertainties.

rank $[I - BB^+] = n - m$ and $[I - BB^+] B = 0$; therefore, the columns of $B^\perp$ can be formed by taking the linearly independent columns of $[I - BB^+]$. Thus, let $\gamma(x,t) \in \mathbb{R}^m$ and $\mu(x,t) \in \mathbb{R}^{n-m}$ be the vector defined by the formula

$$\begin{bmatrix} \gamma(x,t) \\ \mu(x,t) \end{bmatrix} = [B \ B^\perp]^{-1} \phi(x,t)$$

Thus, (2.7) takes the following form:

$$\dot{x} = Ax + B(u_1 + u_0) + B\gamma + B^\perp \mu \qquad (2.10)$$

Then selecting s as in (2.8), we have

$$\dot{s} = GB(u_1 + \gamma) + GB^\perp \mu$$

The control part u_1 should be designed as in (2.9) if GB is positive definite; otherwise, it should be designed as in (2.6). In both cases the condition $M \geq \gamma^+ + (GB)^+ GB^\perp \mu$. Following the equivalent control method, we have that the equivalent control taken from $\dot{s} = 0$ is given by the equation

$$u_{1_{eq}} = -\gamma - (GB)^{-1} GB^\perp \mu$$

Substituting $u_{1_{eq}}$ in (2.10) yields the sliding motion equation:

$$\dot{x} = Ax + Bu_0 + \left[I - B(GB)^{-1} G\right] B^\perp \mu$$

Let us define $\bar{d} := \left[I - B(GB)^{-1} G\right] B^\perp \mu$. Taking $G = B^T$ or $G = B^+$, we get $\bar{d} = B^\perp \mu$, that is, the sliding mode control does not affect the unmatched disturbance part.

Now the question is if by selecting G properly, then the norm of $\bar{d}$ can be made less than the norm of $B^\perp \mu$.

Proposition 2.1. *Let $\bar{\mathcal{G}}$ be the set of matrices*
$$\bar{\mathcal{G}} = \{G \in \mathbb{R}^{m \times n} : \det GB \neq 0\}$$
The optimization problem
$$G^* = \arg\min_{G \in \bar{\mathcal{G}}} \left\| \left[I - B(GB)^{-1} G\right] B^\perp \mu \right\|$$
for $\mu \neq 0$ has as solutions the set of matrices $\{G = QB^T : Q \in \mathbb{R}^{m \times m}$ and $\det Q \neq 0\}$.

Proof. Since $B^\perp \mu$ and $B(GB)^{-1} GB^\perp \mu$ are orthogonal vectors, the norm of $\left\| \left[I - B(GB)^{-1} G\right] B^\perp \mu \right\|$ is always greater than $\|B^\perp \mu\|$. Indeed,
$$\left\| \left[I - B(GB)^{-1} G\right] B^\perp \mu \right\|^2 = \|B^\perp \mu\|^2 + \left\| B(GB)^{-1} GB^\perp \mu \right\|^2$$
That is,
$$\left\| \left[I - B(GB)^{-1} G\right] B^\perp \mu \right\| \geq \|B^\perp \mu\| \tag{2.11}$$
Evidently, if identity (2.11) is achieved, then the norm of
$$\left\| \left[I - B(GB)^{-1} G\right] B^\perp \mu \right\|$$
is minimized with respect to G. The identity is obtained if and only if $B(GB)^{-1} GB^\perp \mu = 0$. Or equivalently, since $\operatorname{rank} B = m$, $GB^\perp \mu = 0$, i.e., $G = QB^T$, where Q is nonsingular. □

□

Notice that the control law is not modified in order to optimize the effect of the unmatched uncertainties, and moreover, an optimal solution G^* is quite simple. The simplest choice is $G^* = B^T$, but $B^+ = (B^T B)^{-1} B^T$ is also another possibility, which moreover facilitates the sliding surface design.

Proposition 2.2. *For an optimal matrix G^*, the Euclidean norm of the disturbance is not amplified, that is,*
$$\|\phi(t)\| \geq \left\| \left[I - B(G^* B)^{-1} G^*\right] B^\perp \mu(t) \right\| \tag{2.12}$$

Proof. From Proposition 2.1, we have that
$$\left\| \left[I - B(G^* B)^{-1} G^*\right] B^\perp \mu(t) \right\| = \left\| \left[I - BB^+\right] B^\perp \mu(t) \right\| = \|B^\perp \mu(t)\| \tag{2.13}$$
Now, since $\phi(t) = B\gamma + B^\perp \mu$ and $B^T B^\perp = 0$, we obtain the equation
$$\|\phi(t)\|^2 = \|B\gamma(t) + B^\perp \mu(t)\|^2 = \|B\gamma(t)\|^2 + \|B\mu(t)\|^2 \geq \|B\mu(t)\|^2 \tag{2.14}$$
Hence, comparing (2.13) and (2.14), we can obtain (2.12). □

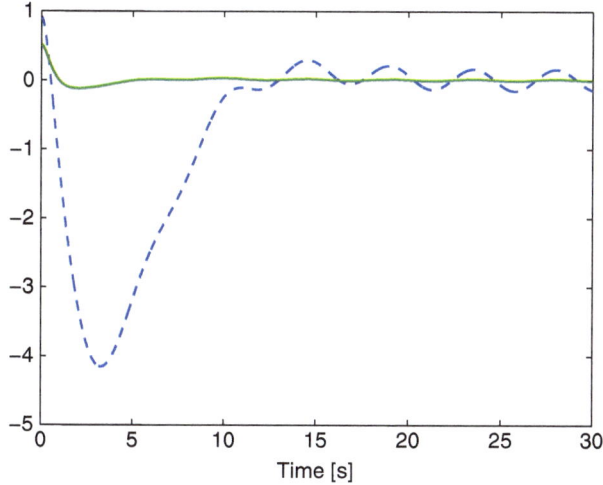

Fig. 2.3. States x_1 (*dashed*) x_2 and (*solid*) using $G = B^T$.

2.6.2 Example: ISM and Unmatched Disturbances

Let us consider the same system as in Sect. 2.6, except that here we consider the unmatched disturbance $\phi = \begin{bmatrix} 0 & 0 & \gamma & 0.1\sin(1.4t) \end{bmatrix}$ and γ is the same function used in Sect. 2.6. The control law is exactly as in Sect. 2.6, except for the choice of matrix G, which according to Proposition 2.1, G is optimal if $G = B^T = \begin{bmatrix} 0 & 0 & 0.19 & 0.14 \end{bmatrix}$. In Sect. 2.6 the goal was simplicity. The argument given in this example revolves around optimality. States x_1 and x_2 are depicted in Fig. 2.3; there we can see that the uncertainties do not affect the trajectories of the system. Figure 2.4 shows the state trajectories for a not optimal G. There we can see that an optimal G does not diminish the effect of the unmatched uncertainties. To compare the effect of the ISM, even in the presence of unmatched disturbances, Fig. 2.5 shows the trajectories of x_1 and x_2 when the ISM control part is omitted ($u = u_0$). It is clear that in this case, the disturbances considerably affect the system; compared with the trajectories of Fig. 2.3 we can see that a well-designed ISM control (with an optimal G) considerably reduces the effect of the disturbances.

2.7 Conclusions

In this section we have seen that the ISM allows to compensate the matched uncertainties. Thus, the performance of the control is equivalent to that of the nominal control which is designed for the nominal system. Furthermore, we have seen that by choosing correctly the matrix projection in the sliding surface the unmatched uncertainties are not increasing in the sliding mode.

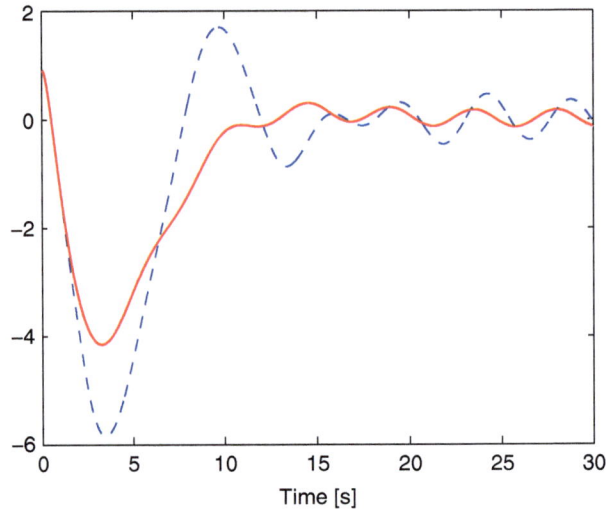

Fig. 2.4. Trajectories of the position for $G = B^T$ (*solid*) and $G = \begin{bmatrix} 0 & 0 & 10 & 0 \end{bmatrix}$ (*dashed*).

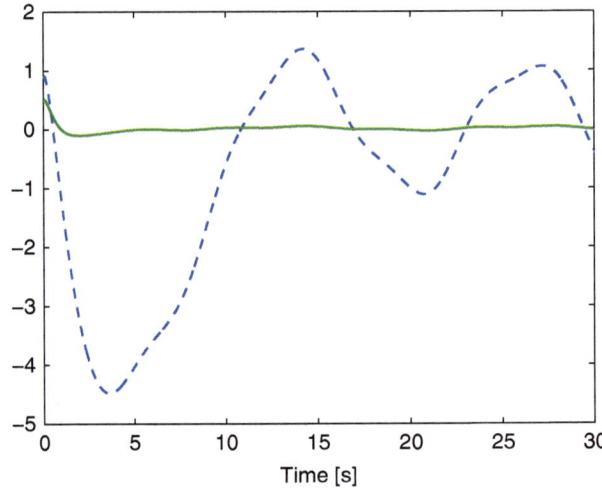

Fig. 2.5. States x_1 (*dashed*) and x_2 (*solid*), without using ISM control.

3

Observer Based on ISM

Abstract In this chapter the concept of output ISM observer for systems with matched unknown inputs is developed. It is shown that using the output as a sliding mode surface one can compensate the unknown inputs. Then if the number of the inputs is more than the number of unknown outputs it is still possible to observe the system. Moreover, the main advantage of such observers is that they can provide, theoretically, an exact value of the state variables right after the initial time moment.

3.1 Motivation

When only the output of a system is available, there are two possibilities for sliding mode control design. One is to use an output feedback control, i.e., design a sliding surface using the output of the system in such a way that the dynamics of the system, during the corresponding sliding motion, have a property required by the designer. This kind of controls can be seen in [41, 42]. Another possibility is to design an observer. To construct an estimator, providing convergence of the generated estimates to the real states, the corresponding sliding surface should be specially designed. There are two main methods for designing sliding mode observers:

- one is aimed to get a zero tracking error between the outputs of the plant and the observer to be constructed (see, e.g., [12, 13, 43, 44]);
- the other one is to design several sliding surfaces to estimate the state step-by-step (see [45, 46]).

Here we design a hierarchical observer which differs from the observers studied in [45–47]. We obtained a vector which is the result of multiplying an observability matrix by the state. Thus, at each k-level of the hierarchy, we estimate a subblock of such vector and so on until we obtain all the vector previously mentioned. The aim is to design an observer whose convergence error can be modified by modifying the accuracy of the sensors and computational resources without modifying the output injection gains. We will show that the observation error can be made arbitrarily small after an arbitrary small time just by adjusting the parameters of the filter required during the realization.

3.2 System Description

Let us consider a linear time invariant system

$$\begin{aligned} \dot{x}(t) &= Ax(t) + Bu(t), \quad x(0) = x_0 \\ y(t) &= Cx(t) \end{aligned} \tag{3.1}$$

where $x(t) \in \mathbb{R}^n$ is the state vector, $u(t) \in \mathbb{R}^m$ is the control law, and $y(t) \in \mathbb{R}^p$ ($1 \leq p < n$) is the output of the system. The pair $\{u(t), y(t)\}$ is assumed to be measurable (available) for all time $t \geq 0$. The current state $x(t)$ and the initial state x_0 are supposed to be non-available. A, B, C are known matrices of appropriate dimension with rank $B = m$ and rank $C = p$. All the solutions of the dynamic systems are defined in Filippov's sense [15]. It is assumed that the pair (A, C) is observable.
We will assume that the following conditions are satisfied:

A4.1. the vector x_0 is supposed to be unknown but belonging to a given ball, that is,

$$\|x_0\| \leq \mu \tag{3.2}$$

A4.2. rank $(CB) = m$

3.3 Observer Design

The hierarchical observer will be based on the reconstruction of vectors $Cx(t)$, $CAx(t)$, and so on until obtaining $CA^{l-1}x(t)$. After arranging the vectors $CA^i x(t)$, we will have obtained the vector $f(t) := \mathcal{O}x(t)$, where

$$\mathcal{O} = \begin{bmatrix} C \\ CA \\ \vdots \\ CA^{l-1} \end{bmatrix}, \quad \mathcal{O} \in \mathbb{R}^{pl \times n} \tag{3.3}$$

By definition l (the observability index) is the least positive integer such that rank $(\mathcal{O}) = n$. Since (A, C) is observable, such an index l always exists (see, e.g., [48]). Hence, to reconstruct $x(t)$, we only need to reconstruct $f(t)$ and then to solve the set of algebraic equations $f(t) = \mathcal{O}x(t)$.
Let $\tilde{x}(t)$ be defined by the following dynamic equation:

$$\dot{\tilde{x}}(t) = A\tilde{x}(t) + Bu(t) + L(y(t) - C\tilde{x}(t)) \tag{3.4}$$

where L must be designed such that the eigenvalues of $\hat{A} := (A - LC)$ have negative real part.

Define $r(t) = x(t) - \tilde{x}(t)$. From (3.1) and (3.4), the dynamic equations governing $r(t)$ are

$$\dot{r}(t) = [A - LC]\, r(t) = \hat{A}r(t) \tag{3.5}$$

Since the eigenvalues of $\hat{A}$ have negative real part, (3.5) is exponentially stable, i.e., there exist constants $\gamma, \eta > 0$ such that

$$\|r(t)\| \leq \gamma e^{-\eta t} \|r(0)\| \leq \gamma e^{-\eta t} (\mu + \|\tilde{x}(0)\|) \tag{3.6}$$

3.3.1 Auxiliary Dynamic Systems and Output Injections

The main goal in the design of the observer is to recover the vectors

$$CA^i x(t), \quad i = \overline{1, l-1}$$

where l is defined as the observability index (see, e.g., [48]). Firstly, to recover $CAx(t)$, let us introduce an *auxiliary state vector* $x_{\mathrm{a}}^{(1)}(t)$ governed by

$$\dot{x}_{\mathrm{a}}^{(1)}(t) = A\tilde{x}(t) + Bu + \tilde{L}\left(C\tilde{L}\right)^{-1} v^{(1)}(t) \tag{3.7}$$

where $\tilde{L}$ satisfies $\det C\tilde{L} \neq 0$ and $x_{\mathrm{a}}^{(1)}(0)$ satisfies

$$Cx_{\mathrm{a}}^{(1)}(0) = y(0)$$

For the variable $s^{(1)} \in \mathbb{R}^p$ defined by

$$s^{(1)}\left(y(t), x_{\mathrm{a}}^{(1)}(t)\right) = Cx(t) - Cx_{\mathrm{a}}^{(1)}(t) \tag{3.8}$$

we have

$$\dot{s}^{(1)}(t) = CA\left(x(t) - \tilde{x}(t)\right) - v^{(1)}(t) \tag{3.9}$$

with the output injection $v^{(1)}(t)$ given by

$$v^{(1)} = \begin{cases} M_1(t) \dfrac{s^{(1)}}{\|s^{(1)}\|} & \text{if } s^{(1)} \neq 0 \\ 0 & \text{if } s^{(1)} = 0 \end{cases}$$

Here the gain scalar function $M_1(t)$ should satisfy the condition

$$M_1(t) > \|CA\|\, \|x - \tilde{x}\| \tag{3.10}$$

to obtain the sliding mode regime. From (3.6), the scalar function $M_1(t)$ may be chosen, for example, in the following manner:

$$M_1(t) = \|CA\|\, [\gamma \exp(-\eta t)(\mu + \|\tilde{x}(0)\|)] + \lambda, \quad \lambda > 0$$

3 Observer Based on ISM

Then, using the Lyapunov function $V = (s, s)$, we obtain

$$s^{(1)}(t) = 0, \ \dot{s}^{(1)}(t) = 0 \ \forall t \geq 0 \tag{3.11}$$

Thus, in view of (3.11) and (3.8), we have

$$Cx(t) = Cx_{\text{a}}^{(1)}(t) \tag{3.12}$$

and from (3.11) and (3.9), the equivalent output injection is

$$v_{\text{eq}}^{(1)}(t) = CAx(t) - CA\tilde{x}(t), \ \forall t > 0$$

Thus, $CAx(t)$ is reconstructed by means of the following representation:

$$CAx(t) = CA\tilde{x}(t) + v_{\text{eq}}^{(1)}(t), \ \forall t > 0 \tag{3.13}$$

The reconstruction of $CAx(t)$ in the form it is expressed in (3.13) is not realizable since $v_{\text{eq}}^{(1)}(t)$ is not directly available. Thereby, below in Sect. 3.4 we explain a method to carry out the estimation of $v_{\text{eq}}^{(1)}(t)$ by means of a first-order low-pass filter applied to $v^{(1)}(t)$.

The next step is to reconstruct the vector $CA^2 x(t)$. To do that, let us design the second auxiliary state vector $x_{\text{a}}^{(2)}(t)$ generated by

$$\dot{x}_{\text{a}}^{(2)}(t) = A^2 \tilde{x}(t) + ABu(t) + \tilde{L}\left(C\tilde{L}\right)^{-1} v^{(2)}(t)$$

where $x_{\text{a}}^{(2)}(0)$ satisfies

$$v_{\text{eq}}^{(1)}(0) + CA\tilde{x}(0) - Cx_{\text{a}}^{(2)}(0) = 0$$

Again, for $s^{(2)} \in \mathbb{R}^p$ defined by

$$s^{(2)}\left(v_{\text{eq}}^{(1)}(t), x_{\text{a}}^{(2)}(t)\right) = CA\tilde{x}(t) + v_{\text{eq}}^{(1)}(t) - Cx_{\text{a}}^{(2)}(t)$$

in view of (3.13), it follows that

$$s^{(2)}\left(v_{\text{eq}}^{(1)}(t), x_{\text{a}}^{(2)}(t)\right) = CAx(t) - Cx_{\text{a}}^{(2)}(t) \tag{3.14}$$

and hence, the time derivative of $s^{(2)}$ is

$$\dot{s}^{(2)}(t) = CA^2(x(t) - \tilde{x}(t)) - v^{(2)}(t) \tag{3.15}$$

Take the output injection $v^{(2)}(t)$ as

$$v^{(2)} = \begin{cases} M_2(t) \dfrac{s^{(2)}}{\|s^{(2)}\|} & \text{if } s^{(2)} \neq 0 \\ 0 & \text{if } s^{(2)} = 0 \end{cases} \tag{3.16}$$

$$M_2(t) > \|CA^2\| \|x - \tilde{x}\|$$

where, by (3.6), $M_2(t)$ given by means of the following formula:

$$M_2(t) = \|CA^2\| \left[\gamma \exp(-\eta t)(\mu + \|\tilde{x}(0)\|)\right] + \lambda, \quad \lambda > 0$$

satisfies (3.16). Thus, following a standard method for proving the existence of the integral sliding mode (Chap. 2), we obtain that

$$s^{(2)}(t) = \dot{s}^{(2)}(t) = 0 \tag{3.17}$$

From (3.17) and (3.15) the equivalent output injection $v_{eq}^{(2)}(t)$ may be represented as

$$v_{eq}^{(2)}(t) = CA^2(x(t) - \tilde{x}(t))$$

and the vector $CA^2 x(t)$ can be recovered by means of the equality:

$$CA^2 x(t) = CA^2 \tilde{x}(t) + v_{eq}^{(2)}(t), \quad t > 0 \tag{3.18}$$

Thus, iterating the same procedure, all the vectors $CA^k x$ can be reconstructed. The abovementioned procedure could be summarized as follows:

(a) *The dynamics of the auxiliary state* $x_a^{(k)}(t)$ *at the kth level are governed by*

$$\dot{x}_a^{(k)}(t) = A^k \tilde{x}(t) + A^{k-1} B u(t) + \tilde{L}\left(C\tilde{L}\right)^{-1} v^{(k)} \tag{3.19}$$

where $\tilde{L} \in \mathbb{R}^{n \times p}$ is a matrix such that $\det\left(C\tilde{L}\right) \neq 0$ for all k. Furthermore, *the output injection* $v^{(k)}$ *at the kth level is*

$$v^{(k)} = \begin{cases} M_k(t) \dfrac{s^{(k)}}{\|s^{(k)}\|} & \text{if } s^{(k)} \neq 0 \\ 0 & \text{if } s^{(k)} = 0 \end{cases}$$
$$M_k(t) > \|CA\| \, \|x(t) - \tilde{x}(t)\|$$
$$M_k(t) \text{ is selected as } M_k(t) = \|CA^k\| \left[\gamma \exp(-\eta t)(\mu + \|\tilde{x}^0\|)\right] + \lambda, \, \lambda > 0. \tag{3.20}$$

(b) Define *the sliding surface* $s^{(k)}$ *at the k-level of the hierarchy* as

$$s^{(k)} = \begin{cases} y - Cx_a^{(1)} & \text{for } k = 1 \\ v_{eq}^{(k-1)} + CA^{k-1}\tilde{x} - Cx_a^{(k)} & \text{for } k > 1 \end{cases} \tag{3.21}$$

where $v_{eq}^{(k-1)}$ is the *equivalent output injection* whose general expression will be obtained in the lemma below and $v_{eq}^{(k-1)}(0)$ and $s^{(k)}(0)$ should satisfy

$$s^{(k)}(0) = \begin{cases} Cy(0) - Cx_a^{(1)}(0) = 0 & \text{for } k = 1 \\ v_{eq}^{(k-1)}(0) + CA^{k-1}\tilde{x}(0) - Cx_a^{(k)}(0) = 0 & \text{for } k > 1 \end{cases} \tag{3.22}$$

Here, $v^{(k)}(t)$ is treated as a *sliding mode* output injection. The equivalent output injection of $v_{eq}^{(k)}(t)$ is given in the next lemma.

Lemma 3.1. *If the auxiliary state vector $x_a^{(k)}$ and the variable $s^{(k)}$ are designed as in (3.19) and (3.21), respectively, then, for all $t \geq 0$,*

$$v_{eq}^{(k)}(t) = CA^k x(t) - CA^k \tilde{x}(t) \tag{3.23}$$

and each $k = \overline{1, l-1}$.

Proof. As it was shown before, the following identity holds:

$$v_{eq}^{(1)}(t) = CAx(t) - CA\tilde{x}(t), \quad \forall t > 0$$

Now, suppose that the equivalent output injection $v_{eq}^{(k-1)}$ is as in (3.23). Thus, substitution of $v_{eq}^{(k-1)}$ in (3.21) gives

$$s^{(k)}\left(v_{eq}^{(k-1)}(t), x_a^{(k)}(t)\right) = CA^{k-1}x(t) - Cx_a^{(k)}(t) \tag{3.24}$$

Differentiating (3.24) yields

$$\dot{s}^{(k)}\left(v_{eq}^{(k-1)}(t), x_a^{(k)}(t)\right) = CA^k\left(x(t) - \tilde{x}(t)\right) - v^{(k)}(t) \tag{3.25}$$

Thus, selecting the Lyapunov function $V = \frac{1}{2}\left\|s^{(k)}\right\|^2$ and $v^{(k)}(t)$ as in (3.20), for any $t \geq 0$, one gets

$$s^{(k)}(t) \equiv 0, \quad \dot{s}^{(k)}(t) \equiv 0 \tag{3.26}$$

Therefore, from (3.26) and (3.25), it follows that

$$v_{eq}^{(k)}(t) \equiv CA^k x(t) - CA^k \tilde{x}(t)$$

The lemma is proven. □

3.4 Observer in the Algebraic Form

From (3.12) and (3.23), we have the following set of equations:

$$\begin{aligned} Cx(t) &= C\tilde{x}(t) + Cx_a^{(1)} - C\tilde{x}(t) \\ CAx(t) &= CA\tilde{x}(t) + v_{eq}^{(1)} \\ &\vdots \\ CA^{l-1}x(t) &= CA^{l-1}\tilde{x}(t) + v_{eq}^{(l-1)} \end{aligned} \tag{3.27}$$

3.4 Observer in the Algebraic Form

or, in a matrix representation

$$\mathcal{O}x(t) = \mathcal{O}\tilde{x}(t) + v_{\text{eq}}(t), \quad \forall t > 0 \tag{3.28}$$

where

$$\mathcal{O} = \begin{bmatrix} C \\ CA \\ \vdots \\ CA^{l-1} \end{bmatrix}, \quad v_{\text{eq}} = \begin{bmatrix} Cx_a^{(1)} - C\tilde{x}(t) \\ v_{\text{eq}}^{(1)} \\ \vdots \\ v_{\text{eq}}^{(l-1)} \end{bmatrix} \tag{3.29}$$

Thus, the left multiplication of (3.28) by $\mathcal{O}^+ := \left[\mathcal{O}^T\mathcal{O}\right]^{-1}\mathcal{O}^T$ implies

$$x(t) \equiv \tilde{x}(t) + \mathcal{O}^+ v_{\text{eq}}(t), \quad \forall t > 0 \tag{3.30}$$

That is why an observer based on the *hierarchical ISM* can be suggested as follows:

$$\hat{x}(t) := \tilde{x}(t) + \mathcal{O}^+ v_{\text{eq}}(t) \tag{3.31}$$

Remark 3.1. Notice that, in general,

$$x^* := \arg\min_{x \in \mathbb{R}^n} \|f - \mathcal{O}x\|^2 = \mathcal{O}^+ f, \quad f \in \mathbb{R}^n$$

where the limit $\mathcal{O}^+ = \lim_{\delta \to 0} \left(\delta^2 I + \mathcal{O}^T\mathcal{O}\right)^{-1}\mathcal{O}^T$ always exists (see [49]) and, moreover,

$$\|f - \mathcal{O}x^*\|^2 = \|(I - \mathcal{O}\mathcal{O}^+)f\|^2$$

This norm is not necessarily equal to zero. In the particular case when $f = \mathcal{O}x$, one has

$$\min_{z \in \mathbb{R}^n} \|f - \mathcal{O}z\|^2 = \|f - \mathcal{O}x^*\|^2 = \|(I - \mathcal{O}\mathcal{O}^+)f\|^2 =$$

$$\|(I - \mathcal{O}\mathcal{O}^+)\mathcal{O}x\|^2 = \|(\mathcal{O} - \mathcal{O}\mathcal{O}^+\mathcal{O})x\|^2 = 0$$

Now we are ready to formulate the main result of this chapter.

Theorem 3.1. *Under the assumptions A4.1–A4.2 and supposing the ideal output integral sliding mode exists, the following identity holds:*

$$\hat{x}(t) \equiv x(t) \quad \forall t > 0 \tag{3.32}$$

Proof. It follows directly from (3.30) and (3.31).

□

Remark 3.2. The realization of the observer (3.31) requires filters whose parameters affect the convergence time of the observer.

3.5 Observer Realization

To carry out the observer in the form (3.31), the surface $s^{(k)}$ must be realizable. Thus, to guarantee the realization of $s^{(k)}$, the equivalent output injection $v_{eq}^{(k)}$ must be available. However, the nonidealities in the implementation of $v^{(k)}$ cause the so-called chattering phenomenon. Thus, we will have a high-frequency signal and therefore $v_{eq}^{(k)}$ cannot be directly obtained from $v^{(k)}$. Nevertheless, $v_{eq}^{(k)}$ could be computed via filtration. Namely, the first-order low-pass filter

$$\tau \dot{v}_{av}^{(k)}(t) + v_{av}^{(k)}(t) = v^{(k)}(t); \quad v_{av}^{(k)}(0) = 0 \tag{3.33}$$

gives an approach of $v_{eq}^{(k)}$ (see Appendix A and [11]). Or, in other words,

$$\lim_{\substack{\tau \to 0 \\ \Delta/\tau \to 0}} v_{av}^{(k)}(t) = v_{eq}^{(k)}(t), t > 0$$

where Δ is proportional to the sampling time (the time that $v^{\alpha,k}$ takes to pass from one state (M) to another $(-M)$). So, selecting $\tau = \Delta^{\eta}$ $(0 < \eta < 1)$, we have the following conditions to realize the OISM observer:

1. use a very small sampling interval Δ;
2. substitute $v_{eq}^{(k-1)}(t)$ in (3.21) by $v_{av}^{(k-1)}(t)$;
3. substitute $v_{eq}^{(k-1)}(0)$ in (3.22) by $v_{av}^{(k-1)}(0) \equiv 0$, i.e., the initial conditions $x_a^{(k)}(0)$ should satisfy the equations

$$CA^{k-1}x_a^{(k-1)}(0) - Cx_a^{(k)}(0) = 0 \text{ for } k > 1$$
$$Cy(0) - Cx_a^{(1)}(0) = 0 \text{ for } k = 1$$

So, the realization of the observer in (3.31) takes the form

$$\hat{x}(t) := \tilde{x}(t) + \mathcal{O}^+ v_{av}(t)$$
$$v_{av} = \left[\left(Cx_a^{(1)} - C\tilde{x}(t) \right)^T \left(v_{av}^{(1)} \right)^T \cdots \left(v_{av}^{(l-1)} \right)^T \right]^T \tag{3.34}$$

An example of the proposed observer design is given in Chap. 4.

3.6 Example

To illustrate the procedure given above, let us take again the linearized model of an inverted pendulum over an inverted cart–pendulum (Fig. 3.1). The motion equations are as follows:

$$\begin{aligned} \dot{x}(t) &= Ax(t) + B(u_0 + u_1) + B\gamma(x,t) \\ y(t) &= Cx(t) \end{aligned} \tag{3.35}$$

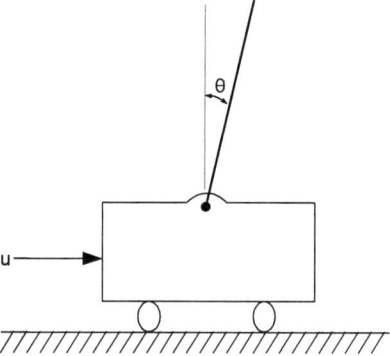

Fig. 3.1. Inverted cart–pendulum.

$$A = \begin{bmatrix} 0 & 0 & 1 & 0 \\ 0 & 0 & 0 & 1 \\ 0 & 1.2586 & 0 & 0 \\ 0 & 7.5514 & 0 & 0 \end{bmatrix}, \; B = \begin{bmatrix} 0 \\ 0 \\ 0.1905 \\ 0.1429 \end{bmatrix}, \; C = \begin{bmatrix} 1 & 0 & 0 & 0 \\ 0 & 0 & 0 & 1 \end{bmatrix}$$

$$\gamma(t) = \begin{cases} -0.4 & n-5 \leq t < n-2.5 \\ 0.4 & n-2.5 \leq t < n \end{cases}, \; n = 5, 10, \ldots$$

The state vector x consists of four state variables: x_1 is the distance between a reference point and the center of inertia of the cart; x_2 represents the angle between the vertical and the pendulum; x_3 represents the linear velocity of the cart; finally, we have that x_4 is equal to the angular velocity of the pendulum. As can be verified, the pair (A, C) has no invariant zeros.

The initial conditions are considered as $x(0) = \begin{bmatrix} 0.3 & 0.2 & 0.1 & -0.1 \end{bmatrix}^T$ and as a consequence we have $y(0) = \begin{bmatrix} 0.3 & -0.1 \end{bmatrix}^T$. As can be verified, the pair $(\tilde{A}, C)$ is observable.

Matrix L chosen so that $(A - LC)$ be Hurwitz. We chose an LQ optimal control with finite horizon, where the estimated state vector is used in place of the original state vector. The simulations were carried out with two sampling steps: $\Delta = 2 \cdot 10^{-5}$ and $\Delta = 2 \cdot 10^{-4}$. In both cases, as the filter constant, the value τ was chosen as $\tau = 150\Delta^{4/5}$.

To realize the suggested observer, the filters suggested in (3.33) must be used. The simulations show that those filters do not affect considerably the observation process (see the observation error $e(t) = x(t) - \hat{x}(t)$ in Figs. 3.2 and 3.3). As we can see in those figures, the convergence to zero is better when Δ is smaller, i.e., the convergence depends only on Δ.

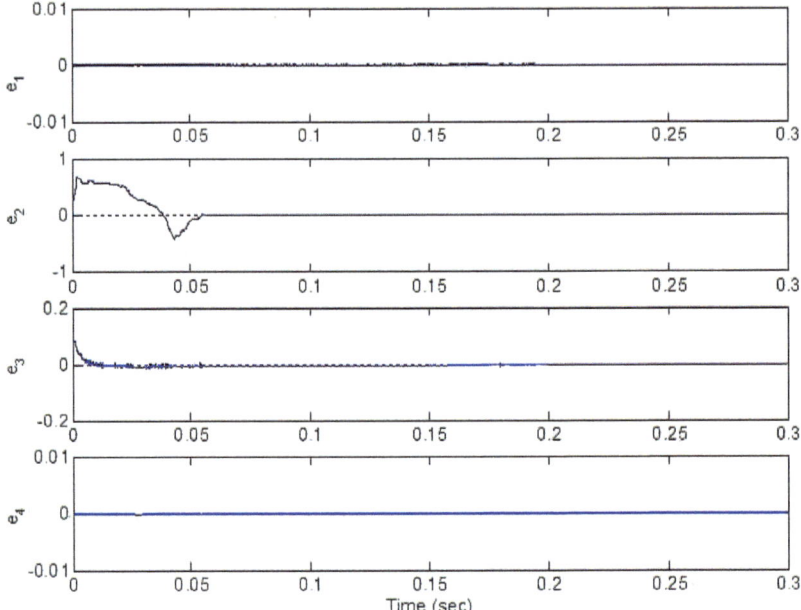

Fig. 3.2. Observation error $e = x - \hat{x}$ using $\Delta = 2 \times 10^{-5}$.

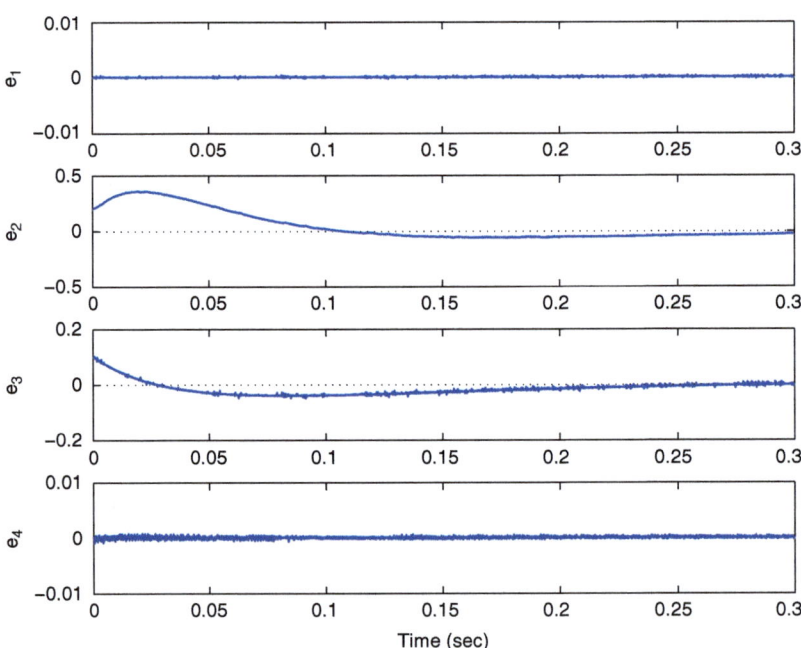

Fig. 3.3. Observation error $e = x - \hat{x}$ using $\Delta = 2 \times 10^{-4}$.

4
Output Integral Sliding Mode Based Control

Abstract Here, the problem of the realization of integral sliding mode controllers based only on output information is discussed. The OISM controller ensures insensitivity of the state trajectory with respect to the matched uncertainties from the initial time moment. In the case when the number of inputs is more than or equal to the number of outputs, the closed-loop system, describing the output integral sliding mode dynamics, is shown to lose observability. For the case when the number of inputs is less than the number of outputs, a hierarchical sliding mode observer is proposed. The realization of the proposed observer requires a filtration to obtain the equivalent output injections. Assigning the first-order low-pass filter parameter small enough (during this filter realization), the convergence time and the observation error can be made arbitrarily small. The results obtained are illustrated by simulations.

4.1 Motivation

The main problem related to the implementation of the ISM consists in the requirement of the *complete knowledge of the state vector,* including the initial one. Obviously, when dealing with *ISM* and only output (no states) information available, it turns out to be *useless* when being applied directly. Here, we present a possible approach to the solution of this problem. We design an ISM controller, using only output information, which compensates the matched uncertainties from the initial time of the control process. It is shown that in the case when the number of inputs is more than (or equal to) the number of outputs, the corresponding ISM dynamics always lose observability and therefore the application of ISM, based only on output information, is useless when the state estimation is required. Then, we use the hierarchical sliding mode observer proposed in Chap. 3.

4.2 System Description

Consider a linear time invariant system with matched disturbances

$$\begin{aligned}\dot{x}(t) &= Ax(t) + Bu(t) + B\gamma(t); \quad x(0) = x^0 \\ y(t) &= Cx(t)\end{aligned} \quad (4.1)$$

where $x(t) \in \mathbb{R}^n$ is the state vector, $u(t) \in \mathbb{R}^m$ is the control law, and $y(t) \in \mathbb{R}^p$ $(1 \leq p < n)$ is the output of the system. The pair $\{u(t), y(t)\}$ is assumed to be measurable (available) for all time $t \geq 0$. The current state $x(t)$ and the initial state x_0 are supposed to be non-available. A, B, C are known matrices of appropriate dimension with rank $B = m$ and rank $C = p$. All the solutions of the dynamic system are defined in Filippov's sense ([15]). We will assume that:

A5.1. the pair (A, B) is controllable and the pair (A, C) is observable;
A5.2. function $\gamma(t)$ is bounded, that is,

$$\|\gamma(t)\| \leq \gamma^+(y, t) \quad (4.2)$$

A5.3. the vector x_0 is supposed to be unknown but belonging to a given ball, that is,

$$\|x_0\| \leq \mu \quad (4.3)$$

A5.4. rank $(CB) = m$.

Let the nominal state be as follows:

$$\dot{x}_0(t) = Ax_0(t) + Bu_0(t), \quad x(0) = x_0 \quad (4.4)$$

Now, for system (4.1), we design the control law u to be

$$u = u_0 + u_1 \quad (4.5)$$

where the control $u_0 \in \mathbb{R}^m$ is the ideal control designed for system (4.4) and $u_1 \in \mathbb{R}^m$ is designed to compensate the matched uncertainty $\phi(t)$ from the initial time.

4.3 OISM Control

This section, firstly, deals with the design of control u_1. Then, a hierarchical integral sliding mode (HISM) observer is applied.

4.4 Output Integral Sliding Modes

Define the auxiliary affine sliding function $s : \mathbb{R}^p \to \mathbb{R}^m$ as follows:

$$s(y) := Gy - \int_0^t [GCA\hat{x}(\tau) + GCBu_0(\tau)] d\tau - Gy(0) \quad (4.6)$$

Here, matrix $G \in \mathbb{R}^{m \times p}$ must satisfy the condition

$$\det(GCB) \neq 0$$

Thus, for the time derivative $\dot{s}$, we have

$$\dot{s} = GCA(x - \hat{x}) + GCB(u_1 + \gamma), \; s(0) = 0 \quad (4.7)$$

Vector $\hat{x}$ represents an observer that will be designed below. We propose the control u_1 in the following form:

$$u_1 = -\beta(t) D^{-1} \frac{s(t)}{\|s(t)\|}, \quad D := GCB \quad (4.8)$$

with $M(t)$ being a scalar gain which satisfies the condition

$$\beta(t) - \left(\|D\| \gamma^+(y, t) + \|GCA\| \|x(t) - \hat{x}(t)\| \right) \geq \lambda > 0$$

where λ is a constant. Selecting the Lyapunov function as $V = \frac{1}{2} \|s\|^2$ and in view of (4.8) and (4.2), differentiating V yields

$$\dot{V} = (s, \dot{s}) = \left(s, GCA(x - \hat{x}) - \beta \frac{s}{\|s\|} + D\gamma \right) \leq$$
$$\leq -\|s\| \left(\beta - \|GCA\| \|x - \hat{x}\| - \|D\| \gamma^+ \right) \leq -\|s\| \lambda \leq 0$$

$((s, \dot{s}) := s^T \dot{s})$. This means that V does not increase in time and since $s(0) = 0$, this implies

$$\frac{1}{2} \|s(t)\| = V(s(t)) \leq V(s(0)) = \frac{1}{2} \|s(0)\| = 0$$

Thus, the identities

$$s(t) = \dot{s}(t) = 0 \quad (4.9)$$

hold for all $t \geq 0$, i.e., there is no reaching phase.
From (4.7) and in view of the equality in (4.9), the *equivalent control* is

$$u_{1eq} = -(GCB)^{-1} GCA(x(t) - \hat{x}(t)) - \gamma \quad (4.10)$$

The substitution of u_{1eq} in (4.1) yields the *sliding mode* equations

$$\dot{x}(t) = \tilde{A}x(t) - B(GCB)^{-1}GCA\hat{x}(t) + Bu_0 \\ y(t) = Cx(t) \qquad (4.11)$$

where $\tilde{A}$ is defined as

$$\tilde{A} := \left[I - B(GCB)^{-1}GC\right]A \qquad (4.12)$$

Lemma 4.1. *When the number of outputs is less than or equal to the number of inputs, the matrix $\tilde{A}$ in (4.12) always belongs to the null space of the matrix C, and, consequently, the pair $\left(\tilde{A}, C\right)$ is not observable.*

The proof of Lemma 4.1 is given at the end of this chapter in Appendix 4.A.

Remark 4.1. Lemma 4.1 means that in the case when $p \leq m$, the ISM control using only output information cannot be realized.

The following lemma establishes the condition, in terms of A, B, and C, providing the observability of the pair $\left(\tilde{A}, C\right)$.

Lemma 4.2. *The pair $\left(\tilde{A}, C\right)$ is observable if and only if the triple (A, B, C) has no zeros, i.e.,*

$$\{s \in \mathbb{C} : \text{rank}\,(P(s)) < n + m\} = \emptyset \qquad (4.13)$$

where $P(s)$ is the system matrix defined as

$$P(s) = \begin{bmatrix} sI - A & B \\ -C & 0 \end{bmatrix} \qquad (4.14)$$

A proof of Lemma 4.2 is given in Appendix 4.B of this chapter.

Remark 4.2. Notice that $\tilde{A}$ defined in (4.12) depends on a matrix G, which can be designed in a nonunique form. However, due to Lemma 4.2, the observability of the pair $\left(\tilde{A}, C\right)$ depends only on the matrices A, B, and C. In other words, the design of G does not affect the observability of $\left(\tilde{A}, C\right)$.

4.5 Design of the Observer

Define G as $G = (CB)^+ := \left[(CB)^T(CB)\right]^{-1}(CB)^T$ which is the pseudo-inverse of CB. Substituting G in (4.11) leads to the following expression:

$$\dot{x}(t) = \tilde{A}x(t) + Bu_0 + B(CB)^+ CA\hat{x}(t) \\ y(t) = Cx(t) \qquad (4.15)$$

4.5 Design of the Observer

where matrix $\tilde{A}$ in (4.12) becomes

$$\tilde{A} = \left[I - B\left(CB\right)^{+} C\right] A$$

It is assumed that:

A5.5. The triple (A, B, C) has no zeros ($\left(\tilde{A}, C\right)$ is observable).

We can follow the design of the observer proposed in Chap. 3 with no essential modifications. Next we will summarize the observer design.
Design the following dynamic system:

$$\dot{\tilde{x}}(t) = \tilde{A}\tilde{x}(t) + Bu_0(t) + B\left(CB\right)^{+} CA\hat{x}(t) + L\left(y(t) - C\tilde{x}(t)\right) \qquad (4.16)$$

where L must be designed so that $\hat{A} := (\tilde{A} - LC)$ only has eigenvalues with negative real part.
Let $r(t) = x(t) - \tilde{x}(t)$, and then, from (4.15) and (4.16), the dynamic equations governing $r(t)$ are

$$\dot{r}(t) = \left[\tilde{A} - LC\right] r(t) = \hat{A}r(t) \qquad (4.17)$$

Then, we should find positive constant numbers γ and η so that

$$\|r(t)\| \leq \gamma e^{-\eta t} \left(\mu + \|\tilde{x}(0)\|\right) \qquad (4.18)$$

Design *the dynamics of the auxiliary state $x_a^{(k)}(t)$ at the kth level* as follows:

$$\dot{x}_a^{(k)}(t) = \tilde{A}^k \tilde{x}(t) + \tilde{A}^{k-1} B \left[u_0(t) + (CB)^{+} CA\hat{x}(t)\right] + \tilde{L}\left(C\tilde{L}\right)^{-1} v^{(k)} \qquad (4.19)$$

where $\tilde{L} \in \mathbb{R}^{n \times p}$ is a matrix such that $\det\left(C\tilde{L}\right) \neq 0$. The initial conditions should satisfy the identities

$$C\tilde{A}^{k-1} x_a^{(k-1)}(0) - C x_a^{(k)}(0) = 0 \text{ for } k > 1$$
$$Cy(0) - C x_a^{(1)}(0) = 0 \text{ for } k = 1$$

The output injection $v^{(k)}$ at the kth level is

$$v^{(k)} = \begin{cases} M_k \dfrac{s^{(k)}}{\|s^{(k)}\|} & \text{if } s^{(k)} \neq 0 \\ 0 & \text{if } s^{(k)} = 0 \end{cases} \qquad (4.20)$$
$$M_k \geq \left\|C\tilde{A}^k\right\| \left(\gamma e^{-\eta t} \left(\mu + \|\tilde{x}^0\|\right)\right) + \lambda, \ \lambda > 0$$

Define *the sliding surface $s^{(k)}$ at the kth level of the hierarchy* as

$$s^{(k)}(t) = \begin{cases} y(t) - C x_a^{(1)}(t) & \text{for } k = 1 \\ v_{av}^{(k-1)}(t) + C\tilde{A}^{k-1}\tilde{x}(t) - C x_a^{(k)}(t) & \text{for } k > 1 \end{cases} \qquad (4.21)$$

where $v_{\text{av}}^{(k)}$ is the output of the low-pass filter

$$\tau \dot{v}_{\text{av}}^{(k)}(t) + v_{\text{av}}^{(k)}(t) = v^{(k)}(t); \quad v_{\text{av}}^{(k)}(0) = 0 \tag{4.22}$$

Thus, the *hierarchical ISM* observer takes the following form:

$$\begin{aligned}
\hat{x}(t) &:= \tilde{x}(t) + O^+ v_{\text{av}}(t) \\
O^T &= \begin{bmatrix} C^T & (CA)^T & \cdots & (CA^{l-1})^T \end{bmatrix} \\
v_{\text{av}} &= \begin{bmatrix} \left(Cx_a^{(1)} - C\tilde{x}(t)\right)^T & \left(v_{\text{av}}^{(1)}\right)^T & \cdots & \left(v_{\text{av}}^{(l-1)}\right)^T \end{bmatrix}^T
\end{aligned} \tag{4.23}$$

4.6 LQ Control Law

Here, as a case of study, we design the nominal control u_0 as an optimal control based on the standard LQ index for a finite horizon. The control u_0 is designed for the nominal dynamics, i.e.,

$$\dot{x}(t) = Ax(t) + Bu_0, \ x(0) = x_0 \tag{4.24}$$

Control u_0 is an admissible control (belonging to a set U_{adm} of piecewise continuous functions) which minimizes the following standard LQ index:

$$J_{t_f}(u_0(\cdot)) := x^\top(t_f) F x(t_f) + \int_{t=0}^{t_f} \left(x^\top(t) Q x(t) + u_0^\top(t) R u_0(t) \right) dt$$

where $F = F^\top \geq 0$, $Q = Q^\top \geq 0$, and $R = R^\top > 0$. Thus, the aim of the control u_0 is to minimize the index $J(u(\cdot))$, i.e.,

$$u_0^*(\cdot) = \arg \min_{u_0 \in U_{adm}} J_{t_f}(u_0(\cdot)) \tag{4.25}$$

Thus, the control law solving (4.25) for (4.24) (e.g., see [35]) is of the form

$$u_0^*(x(t)) = -R^{-1} B^\top P(t) x(t)$$

with $P(t) \in \mathbb{R}^{n \times n}$ satisfying the differential Riccati equation

$$\begin{aligned}
\dot{P}(t) + P(t) A + A^\top P(t) - P(t) B R^{-1} B^\top P(t) + Q &= 0 \\
P(t_f) &= F
\end{aligned} \tag{4.26}$$

Since the state x can be estimated with any required accuracy, the estimated state $\hat{x}$ is used to realize the control u_0, i.e., the control u_0 should be designed as

$$u_0(t) = -R^{-1} B^\top P(t) \hat{x}(t) \tag{4.27}$$

with $\hat{x}(t)$ being designed as in (4.23). That is, since we have compensated the matched uncertainties and we can ensure the estimation error being arbitrarily small after an arbitrarily small time, we can design the control u_0 for the nominal system, but being applied to system (4.1).

The proposed OISM algorithm can be summarized as follows:

(1) design matrix L such that the eigenvalues of $\hat{A} := (\tilde{A} - LC)$ have negative real part;
(2) compute the scalar gain $\beta(t)$ as in (4.8);
(3) design the auxiliary systems $x_a^{(k)}$ as in (4.19) with the sliding surfaces $s^{(k)}$ as in (4.21) and compute the constants M_k, $k = 1, .., l - 1$;
(4) run simultaneously the observer $\hat{x}$ according to (4.23) and the controllers u_0, u_1 according to (4.27) and (4.8), respectively.

4.7 Example

To illustrate the procedure given above, let us take again the linearized model of an inverted pendulum over an inverted cart–pendulum (see Fig. 4.1). The control problem is to maintain the inverted pendulum in a vertical line. The control law is the force applied to the trolley. The motion equations are as follows:

$$\dot{x}(t) = Ax(t) + B(u_0 + u_1) + B\gamma(x,t) \\ y(t) = Cx(t) \quad (4.28)$$

$$A = \begin{bmatrix} 0 & 0 & 1 & 0 \\ 0 & 0 & 0 & 1 \\ 0 & 1.2586 & 0 & 0 \\ 0 & 7.5514 & 0 & 0 \end{bmatrix}, B = \begin{bmatrix} 0 \\ 0 \\ 0.1905 \\ 0.1429 \end{bmatrix}, C = \begin{bmatrix} 1 & 0 & 0 & 0 \\ 0 & 0 & 0 & 1 \end{bmatrix}$$

$$\gamma(t) = \begin{cases} -0.4 & n - 5 \leq t < n - 2.5 \\ 0.4 & n - 2.5 \leq t < n \end{cases}, n = 5, 10, \ldots$$

The state vector x consists of four state variables: x_1 is the distance between a reference point and the center of inertia of the trolley; x_2 represents the angle

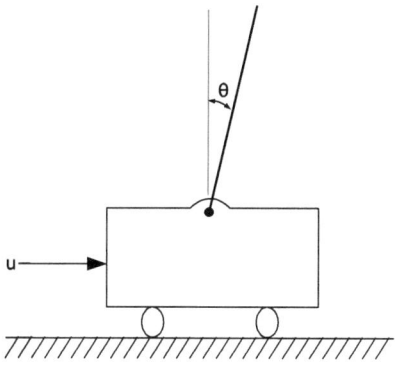

Fig. 4.1. Inverted cart–pendulum.

between the vertical and the pendulum; x_3 represents the linear velocity of the trolley; finally, we have that x_4 is equal to the angular velocity of the pendulum. As can be verified, the pair (A, C) has no invariant zeros. By Lemma 4.2, it implies that $(\tilde{A}, C)$ is observable ($\tilde{A} = \left[I - B\left(CB\right)^{+} C \right] A$).

The initial conditions are considered as $x(0) = \begin{bmatrix} 0.3 & 0.2 & 0.1 & -0.1 \end{bmatrix}^T$, and as a consequence we have $y(0) = \begin{bmatrix} 0.3 & -0.1 \end{bmatrix}^T$. The matrix $\tilde{A}$ takes the form

$$\tilde{A} = \begin{bmatrix} 0 & 0 & 1 & 0 \\ 0 & 0 & 0 & 1 \\ 0 & -8.81 & 0 & 0 \\ 0 & 0 & 0 & 0 \end{bmatrix}$$

As can be verified, the pair $(\tilde{A}, C)$ is observable.
Matrix L was calculated as follows:

$$L = \begin{bmatrix} 4.6234 & -0.3148 \\ -1.3423 & 0.5548 \\ 10.2373 & -1.7542 \\ -0.3148 & 0.9492 \end{bmatrix}$$

The weighing matrices Q, R, and F were chosen as $Q = 20I$, $R = 0.5$, and $F = 20I$.

The simulations were carried out with two sampling steps: $\Delta = 2 \cdot 10^{-5}$ and $\Delta = 2 \cdot 10^{-4}$. In both cases, as the filter constant, the value τ of (4.22) was chosen as $\tau = 150\Delta^{4/5}$. The trajectories of the state vector, when $\hat{x}$ (called xe in the graph) is used in the control u and when x is used in the control u, are depicted in Figs. 4.2 and 4.3.

4.A Proof of Lemma 4.1

Proof. Consider system (4.1) with $p \leq m$ and $rank(CB) = p$. Suppose that the control law u is designed in the following way:

$$u = u_0 + u_1$$

where u_0 is the nominal control used after the compensation of the matched disturbance and u_1 is designed to compensate the matched disturbance. At first we will consider the case when $p = m$ and next the case when $p < m$.

1. Consider the case when $p = m$.

Define the auxiliary function s as follows:

$$s = Gy - \int_0^t GCA\hat{x}(\tau) + GCBu_0(\tau) \, d\tau - Gy(0) \qquad (4.29)$$

4.A Proof of Lemma 4.1

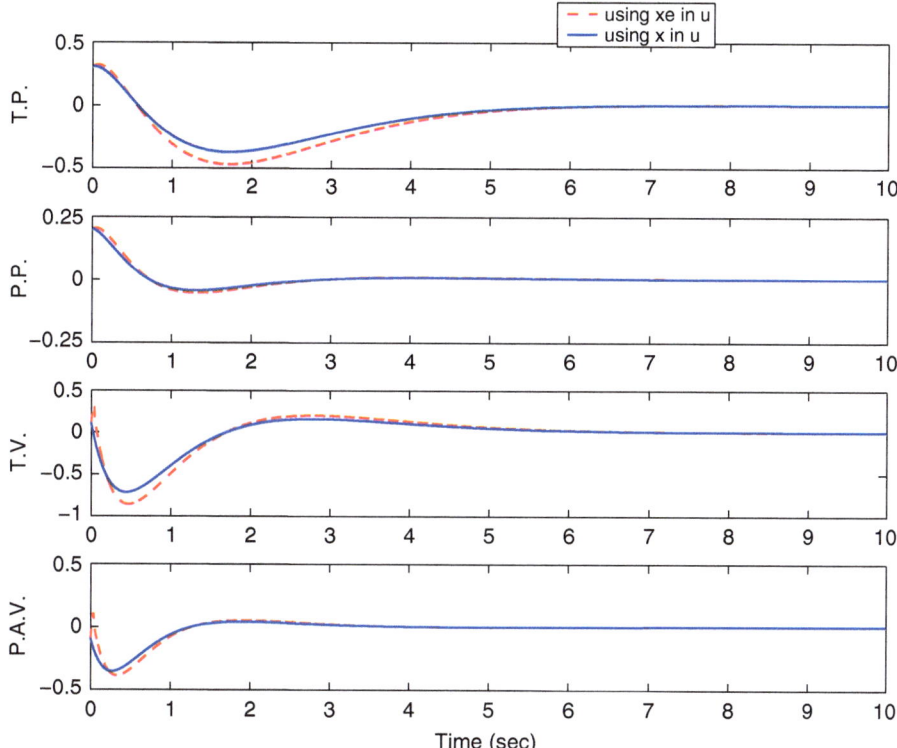

Fig. 4.2. Trajectories of x using $\Delta = 2 \times 10^{-5}$. Trolley position (T.P.), pendulum position (P.P.), trolley velocity (T.V.), and pendulum angular velocity (P.A.V.).

Matrix $G \in \mathbb{R}^{m \times m}$ must satisfy $\text{rank}(GCB) = m$, but this is only satisfied when $\det(G) \neq 0$. Following the same process as in 4.4, one has

$$u_{1\text{eq}} = -(GCB)^{-1} GCA(x - \hat{x}) - \gamma$$

Substitution of $u_{1\text{eq}}$ in system (4.1) yields

$$\dot{x}(t) = \tilde{A}x(t) + B(GCB)^{-1} GCA\hat{x}(t) + Bu_0$$
$$y(t) = Cx(t)$$

Recall that $\tilde{A} = \left[I - B(GCB)^{-1} GC\right] A$. Then by pre-multiplying $\tilde{A}$ by GC one gets

$$GC\tilde{A} = GC\left[I - B(GCB)^{-1} GC\right] A = 0$$

This means $\tilde{A}$ belongs to the null space of GC and since G is a nonsingular matrix, then $\tilde{A}$ belongs to the null space of C and it implies that $(\tilde{A}, C)$ is not observable.

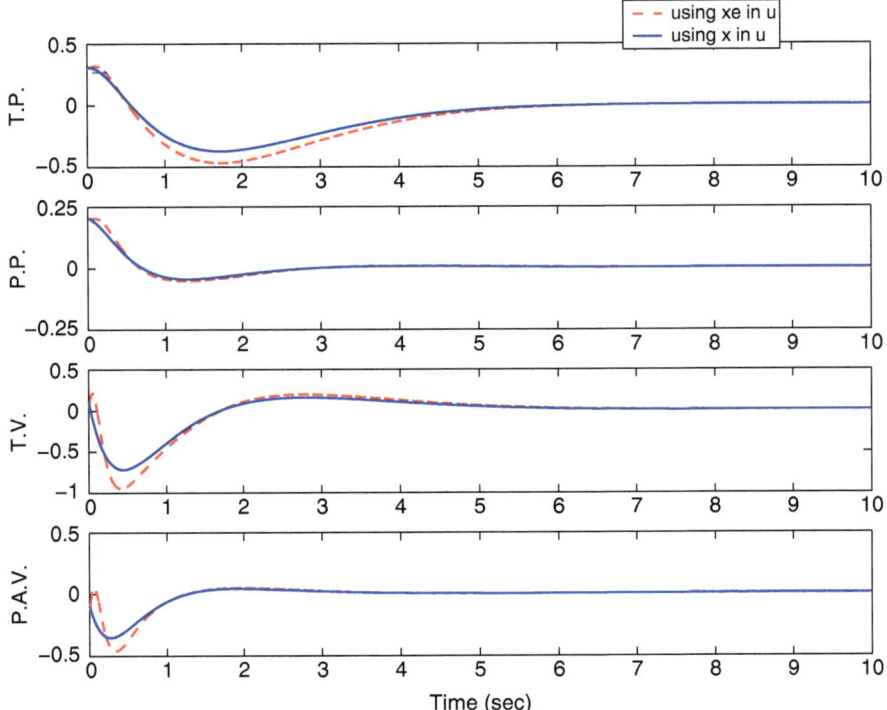

Fig. 4.3. Trajectories of x using $\Delta = 2 \times 10^{-4}$. Trolley position (T.P.), pendulum position (P.P.), trolley velocity (T.V.), and pendulum angular velocity (P.A.V.).

2. Now suppose that $p < m$.

Let the auxiliary function s as in (4.29). Since $\operatorname{rank}(CB) = p$ and $p < m$, then there is no matrix $G \in \mathbb{R}^{m \times p}$ satisfying $\operatorname{rank}(GCB) = m$. That is why the sliding surface s cannot be designed in a space of dimension greater than p. Let us define s in the space $\mathbb{R}^p$, that is,

$$s = Gy - Gy(0) - \int_0^t [GCA\hat{x}(\tau) + GCBu_0(\tau)]\,d\tau$$

where the matrix $G \in \mathbb{R}^{p \times p}$. Thus, the time derivative $\dot{s}$ is as follows:

$$\dot{s} = GC\left[A(x - \hat{x}) + B(u_1 + \gamma)\right]$$

Since in this case $p < m$, there is no matrix G satisfying $\det(GCB) \neq 0$. Hence, to produce the sliding mode, the control u_1 should be designed as $u_1 := \bar{F}\bar{u}$, where the matrix $\bar{F} \in \mathbb{R}^{m \times p}$ should satisfy $\operatorname{rank}(GCB\bar{F}) = p$. Thus $B\bar{F}$ can be considered as the new matrix of input distribution and $\bar{u}$ as the new control. In this form, we can consider that the number of inputs is p, i.e., we have the same number of inputs as the number of outputs. Hence, we can follow the same proof used for the case 1.

□

4.B Proof of Lemma 4.2

Proof. Lemma (4.2) asserts that for every complex scalar s, the equivalence

$$\text{rank} \begin{bmatrix} sI - A & B \\ -C & 0 \end{bmatrix} = n + m \iff \text{rank} \begin{bmatrix} sI - \tilde{A} \\ C \end{bmatrix} = n$$

is fulfilled. Let $B^\perp$ be a matrix so that $B^\perp B = 0$ and $\text{rank}\, B^\perp = n - m$. Define the matrices V and U in the following form:

$$V := \begin{bmatrix} B^\perp \\ (GCB)^{-1} GC \end{bmatrix}, V^{-1} = \begin{bmatrix} \left[I - B(GCB)^{-1} GC\right] B^{\perp +} & B \end{bmatrix}$$

$$U := \begin{bmatrix} (CB)^\perp \\ G \end{bmatrix}, U^{-1} = \begin{bmatrix} \left[I - CB(GCB)^{-1} GC\right](CB)^{\perp +} & CB(GCB)^{-1} \end{bmatrix}$$

Before proving the required equivalence, we need to express the following matrices into an expanded form, i.e.,

$$VAV^{-1} = \begin{bmatrix} A_{11} & A_{12} \\ A_{21} & A_{22} \end{bmatrix}, UCV^{-1} = \begin{bmatrix} C_1 & 0 \\ 0 & GCB \end{bmatrix} \quad (4.30)$$

where $A_{11} \in \mathbb{R}^{n-m \times n-m}$ and $C_1 \in \mathbb{R}^{p-m \times n-m}$. We obtain

$$V\tilde{A}V^{-1} = \begin{bmatrix} A_{11} & A_{12} \\ A_{21} & A_{22} \end{bmatrix} - \begin{bmatrix} 0 \\ I \end{bmatrix} \begin{bmatrix} A_{21} & A_{22} \end{bmatrix} = \begin{bmatrix} A_{11} & A_{12} \\ 0 & 0 \end{bmatrix} \quad (4.31)$$

Then, from (4.30) and (4.31), and since $\det(GCB) \neq 0$, we have the following equivalences:

$$\text{rank} \begin{bmatrix} sI - A & B \\ -C & 0 \end{bmatrix} = n + m \iff \text{rank} \begin{bmatrix} sI - VAV^{-1} & VB \\ -UCV^{-1} & 0 \end{bmatrix} = n + m \iff$$

$$\text{rank} \begin{bmatrix} sI - A_{11} & -A_{12} & 0 \\ -A_{21} & sI - A_{22} & I \\ -C_1 & 0 & 0 \\ 0 & -GCB & 0 \end{bmatrix} = n + m \iff \text{rank} \begin{bmatrix} sI - A_{11} \\ -C_1 \end{bmatrix} = n - m \iff$$

$$\iff \text{rank} \begin{bmatrix} sI - A_{11} & -A_{12} \\ 0 & sI \\ -C_1 & 0 \\ 0 & -GCB \end{bmatrix} = n \iff \text{rank} \begin{bmatrix} sI - V\tilde{A}V^{-1} \\ -UCV^{-1} \end{bmatrix} = n \iff$$

$$\text{rank} \left\{ \begin{bmatrix} I_n & 0 \\ 0 & U \end{bmatrix} \begin{bmatrix} V & 0 \\ 0 & I_p \end{bmatrix} \begin{bmatrix} sI - \tilde{A} \\ -C \end{bmatrix} V^{-1} \right\} = n \iff \text{rank} \begin{bmatrix} sI - \tilde{A} \\ -C \end{bmatrix} = n \quad (4.32)$$

and so this lemma is proven. $\square$

Part II

MIN–MAX OUTPUT ROBUST LQ CONTROL

5
The Robust Maximum Principle

Abstract The purpose of this chapter is to explore the possibilities of the maximum principle (MP) approach for the class of min–max control problems dealing with construction of the optimal control strategies for a class of uncertain models given by a system of ordinary differential equations with unknown parameters from a given finite set. The problem under consideration belongs to the class of optimization problems of the min–max type and consists in the design of a control providing a "good" behavior if applied to all models from a given class. Here a version of the *robust maximum principle* applied to the *min–max Bolza problem* with a terminal set is presented. The cost function contains a terminal term as well as an integral one. A fixed horizon is considered. The main result deals with finite parametric uncertain sets involved in a model description. The *min–max LQ control problem* is considered in detail.

5.1 Min–Max Control Problem in the Bolza Form

The *min–max control* problem, dealing with different classes of partially known nonlinear systems, can be formulated in such a way that

- the operation of the maximization is taken over a set of uncertainty or possible scenarios;
- the operation of the minimization is taken over control strategies within a given set.

5.1.1 System Description

Consider a system of multimodel controlled plants

$$\dot{x} = f^\alpha(x, u, t) \tag{5.1}$$

where

- $x = (x^1, \ldots, x^n)^T \in \mathbb{R}^n$ is its state vector,
- $u = (u^1, \ldots, u^r)^T \in \mathbb{R}^r$ is the control that may run over a given control region $U \subset \mathbb{R}^r$,
- α is a parameter belonging to a given parametric set $\mathcal{A}$ which is assumed to be finite that corresponds to a *multimodel* situation and $t \in [0, T]$.

The usual restrictions are imposed on the right-hand side

$$f^\alpha (x, u, t) = \left(f^{\alpha, 1} (x, u, t), \ldots, f^{\alpha, n} (x, u, t) \right)^T \in \mathbb{R}^n$$

that is,

- the *continuity* with respect to the collection of the arguments x, u, t;
- the *differentiability* (or Lipschitz condition) with respect to x

One more restriction is formulated below.

5.1.2 Feasible and Admissible Control

Remember that a function $u(t)$, $0 \le t \le T$ is said to be a *feasible control* if it is piecewise continuous and $u(t) \in U$ for all $t \in [0, T]$. For convenience, every feasible control is assumed to be right-continuous, that is,

$$u(t) = u(t + 0) \text{ for } 0 \le t < T \tag{5.2}$$

and, moreover, $u(t)$ is continuous at the terminal moment:

$$u(T) = u(T - 0) \tag{5.3}$$

For a given feasible control $u(t)$, $t_0 \le t \le T$, consider the corresponding solution

$$x^\alpha (t) = \left(x^{\alpha, 1} (t), \ldots, x^{\alpha, n} (t) \right)^T$$

of (5.1) with the initial condition

$$x^\alpha (0) = x_0^\alpha$$

Any feasible control $u(t)$, $0 \le t \le T$ as well as all solutions $x^\alpha(t), \alpha \in \mathcal{A}$ are assumed to be defined on the whole segment $[0, T]$ (this is the additional restriction to the right-hand side of (5.1)).

In the space R^n the terminal set $\mathcal{M}$ given by the inequalities

$$g_l(x) \le 0 \ (l = 1, \ldots, L) \tag{5.4}$$

is defined, where $g_l(x)$ is a smooth real function of $x \in \mathbb{R}^n$.

5.1 Min–Max Control Problem in the Bolza Form

For a given feasible control $u(t)$, $0 \leq t \leq T$ we are interested in the corresponding trajectory starting from the initial point x^α. However, the possible realized value of $\alpha \in \mathcal{A}$ is *a priory* unknown. That's why the *family* of trajectories $x^\alpha(t)$ with insufficient information about the realized trajectory is considered.

The control $u(t)$, $0 \leq t \leq T$ is said to be *admissible* or that *it realizes the terminal condition (5.4)*, if it is feasible and for every $\alpha \in \mathcal{A}$ the corresponding trajectory $x^\alpha(t)$ satisfies the inclusion

$$x^\alpha(T) \in \mathcal{M} \tag{5.5}$$

The set of all admissible control strategies will be denoted by $\mathcal{U}_{adm}$.

5.1.3 The Cost Function and the Min–Max Control Problem

Let the cost function in the Bolza form contain an integral term as well as a terminal one, that is,

$$h^\alpha := h_0(x^\alpha(T)) + \int_{t=0}^{t=T} f^{n+1}\left(x^\alpha(t), u(t), t\right) dt \tag{5.6}$$

The end time-point T is assumed to be fixed and $x^\alpha(t) \in \mathbb{R}^n$. Analogously, since the realized value of the parameter α is unknown, the *worst (highest) cost* can be defined as follows:

$$F = \max_{\alpha \in \mathcal{A}} h^\alpha \tag{5.7}$$

The function F depends only on the considered admissible control $u(t)$, $0 \leq t \leq T$. In other words, we wish to construct the admissible control action which provides a "good" behavior for a given collection of models that may be associated with the multimodel robust optimal design.

Definition 5.1. *A control $u(\cdot)$ is said to be **robust optimal** if*

(i) it realizes the terminal condition, that is, it is admissible;
*(ii) it realizes the **minimal** worst (highest) cost F (among all admissible controls).*

Thus the *Robust Optimization Problem* consists in finding a control action $u(t)$, $0 \leq t \leq T$, which realizes

$$\min_{u(\cdot) \in \mathcal{U}_{adm}} F = \min_{u(\cdot) \in \mathcal{U}_{adm}} \max_{\alpha \in \mathcal{A}} h^\alpha \tag{5.8}$$

where the minimum is taken over the set $\mathcal{U}_{adm}$ of all admissible control strategies. This is the *min–max Bolza problem*.

48   5 The Robust Maximum Principle

5.1.4 The Mayer Form Representation

Below we follow the standard transformation scheme. For each fixed $\alpha \in \mathcal{A}$ introduce the $(n+1)$-dimensional space $\mathbb{R}^{n+1}$ of the variables $x = (x_1, \ldots, x_n, x_{n+1})$ where the first n coordinates satisfy (5.1) and the component x_{n+1} is given by

$$x^{\alpha,n+1}(t) := \int_{\tau=0}^{t} f^{n+1}\left(x^{\alpha}(\tau), u(\tau), \tau\right) d\tau$$

or, in the differential form,

$$\dot{x}^{\alpha,n+1}(t) = f^{n+1}\left(x^{\alpha}(t), u(t), t\right) \tag{5.9}$$

with the initial condition for the last component given by

$$x^{\alpha,n+1}(0) = 0$$

As a result, the initial Robust Optimization Problem in Bolza form can be reformulated in the space $\mathbb{R}^{n+1}$ as the *Mayer Problem* (without the integral term) with the cost function

$$h^{\alpha} = h_0(x^{\alpha}(T)) + x^{\alpha,n+1}(T) \tag{5.10}$$

where the function $h_0(x^{\alpha})$ does not depend on the last coordinate $x^{\alpha,n+1}$, that is,

$$\frac{\partial}{\partial x^{\alpha,n+1}} h_0(x^{\alpha}) = 0$$

So, the Mayer Problem with the cost function (5.10) is equivalent to the initial optimization problem (5.8) in the Bolza form.

5.1.5 The Hamiltonian Form

Let

$$\bar{x}^{\alpha}(t) = \left(x^{\alpha,1}(t), \ldots, x^{\alpha,n}(t), x^{\alpha,n+1}(t)\right) \in \mathbb{R}^{n+1}$$

be a solution of systems (5.1) and (5.9). We also introduce for any $\alpha \in \mathcal{A}$ the following *conjugate* (or *adjoint*) variables:

$$\bar{\psi}_{\alpha}(t) = \left(\psi_{\alpha,1}(t), \ldots, \psi_{\alpha,n}(t), \psi_{\alpha,n+1}(t)\right) \in \mathbb{R}^{n+1}$$

satisfying the ODE-system of the adjoint variables:

5.1 Min–Max Control Problem in the Bolza Form

$$\dot{\psi}_{\alpha,i} = -\sum_{k=1}^{n+1} \frac{\partial f^{\alpha,k}(x^\alpha(t), u(t))}{\partial x^{\alpha,i}} \psi_{\alpha,k} \qquad (5.11)$$

with the terminal condition

$$\psi_{\alpha,j}(T) = b_{\alpha,j}, \quad t_0 \leq t \leq T$$

$$\alpha \in \mathcal{A}, \quad j = 1, \ldots, n+1 \qquad (5.12)$$

Let now $\bar{\psi}_\circ = (\psi_{\alpha,i}) \in \mathbb{R}_\circ$ be a covariant vector and

$$\bar{f}^\circ(\bar{x}^\circ, u) = \left(f^{\alpha,k}\right), \quad \bar{x}^\circ = \left(x^{\alpha,k}\right)$$

Introduce the *Hamiltonian function*

$$\mathcal{H}^\circ(\bar{\psi}_\circ, \bar{x}^\circ, u, t) := \langle \bar{\psi}_\circ, \bar{f}^\circ(\bar{x}^\circ, u, t) \rangle =$$

$$\sum_{\alpha \in \mathcal{A}} \langle \bar{\psi}_\alpha, \bar{f}^\alpha(x^\alpha, u, t) \rangle = \sum_{\alpha \in \mathcal{A}} \sum_{i=1}^{n+1} \psi_{\alpha,i} f^{\alpha,i}(x^\alpha, u, t) \qquad (5.13)$$

and remark that $\mathcal{H}^\circ(\bar{\psi}_\circ, \bar{x}^\circ, u)$ is the sum of "usual" Hamiltonian functions:

$$\mathcal{H}^\circ(\bar{\psi}_\circ, \bar{x}^\circ, u, t) = \sum_{\alpha \in \mathcal{A}} \langle \bar{\psi}_\alpha, \bar{f}^\alpha(x^\alpha, u, t) \rangle$$

The function (5.13) allows us to rewrite the conjugate equations (5.11) for the plant (5.1) in the following vector form:

$$\frac{d}{dt}\bar{\psi}_\circ = -\frac{\partial \mathcal{H}^\circ(\bar{\psi}_\circ, \bar{x}^\circ(t), u(t), t)}{\partial \bar{x}^\circ} \qquad (5.14)$$

Let now $b_\circ = (b_{\alpha,i}) \in \bar{\mathbb{R}}_\circ$ be a covariant vector. Denote by $\psi_\circ(t)$ the solution of equation (5.14) with the terminal condition

$$\psi_\circ(T) = b_\circ$$

We say that the control $u(t)$, $t_0 \leq t \leq T$, satisfies the *maximum condition* with respect to the pair $x^\circ(t), \psi_\circ(t)$ if

$$u(t) = \arg\max_{u \in U} \mathcal{H}^\circ(\psi_\circ(t), x^\circ(t), u, t) \quad \forall t \in [t_0, T] \qquad (5.15)$$

that is, $\forall u \in U$, $t \in [t_0, T]$ we have

$$\mathcal{H}^\circ(\psi_\circ(t), x^\circ(t), u(t), t) \geq \mathcal{H}^\circ(\psi_\circ(t), x^\circ(t), u, t)$$

5.2 Robust Maximum Principle

5.2.1 Main Result

Following [9], we may formulate the main result dealing with the necessary conditions for the robust optimality of an admissible control.

Theorem 5.1 (The Maximum Principle for the Bolza Problem with a Terminal Set). *Let $u(t)$ ($t \in [t_0, T]$) be an **admissible control** and $x^\alpha(t)$ be the corresponding solution of (5.1) with the initial condition $x^\alpha(0) = x_0^\alpha$ ($\alpha \in \mathcal{A}$). The parametric uncertainty set $\mathcal{A}$ is assumed to be finite. For robust optimality of a control $u(t)$, $t_0 \leq t \leq T$, it is **necessary** that there exists a vector $b_\circ \in \bar{R}_\circ$ and nonnegative real values $\mu(\alpha)$ and $\nu_l(\alpha)$ ($l = 1, \ldots, L$) defined on $\mathcal{A}$ such that the following conditions are satisfied:*

(i) *the **maximality condition**: denote by $\psi_\circ(t)$, $t_0 \leq t \leq T$ the solution of equation (5.11) with the terminal condition (5.12); then the robust optimal control $u(t)$, $t_0 \leq t \leq T$ satisfies the maximality condition (5.15);*

(ii) *the **complementary slackness conditions**: for every $\alpha \in \mathcal{A}$, either the equality $h^\alpha = F^0$ holds, or $\mu(\alpha) = 0$, that is,*

$$\mu(\alpha)\left(h^\alpha - F^0\right) = 0$$

moreover, for every $\alpha \in \mathcal{A}$, either the equality $g_l(x^\alpha(T)) = 0$ holds, or $\nu_l(\alpha) = 0$, that is,

$$\nu(\alpha) g(x^\alpha(T)) = 0$$

(iii) *the **transversality condition**: for every $\alpha \in \mathcal{A}$, the equalities*

$$\psi_\alpha(T) + \mu(\alpha)\, \mathrm{grad}\, h_0(x^\alpha(T)) + \sum_{l=1}^{L} \nu_l(\alpha)\, \mathrm{grad}\, g_l(x^\alpha(T)) = 0$$

and

$$\psi_{\alpha,n+1}(T) + \mu(\alpha) = 0$$

hold;

(iv) *the **nontriviality condition**: there exists $\alpha \in \mathcal{A}$ such that either $\psi_\alpha(T) \neq 0$, or at least one of the numbers $\mu(\alpha), \nu_l(\alpha)$ is different from zero, that is,*

$$|\psi_\alpha(T)| + \mu(\alpha) + \sum_{l=1}^{L} \nu_l(\alpha) > 0$$

The proof of this theorem is based on the so-called Tent Method and can be found in [50, 51] and [9].

5.3 Min–Max Linear Quadratic Multimodel Control

5.3.1 The Problem Formulation

Consider the following class of nonstationary linear systems given by

$$\begin{cases} \dot{x}^\alpha(t) = A^\alpha(t) x^\alpha(t) + B^\alpha(t) u(t) + d^\alpha(t) \\ x^\alpha(0) = x_0^\alpha \end{cases} \quad (5.16)$$

where $x^\alpha(t)$, $d^\alpha(t) \in R^n$, $u(t) \in R^r$ and the functions $A^\alpha(t)$, $B^\alpha(t)$, $d^\alpha(t)$ are continuous on $t \in [0,T]$. The following performance index is defined as

$$h^\alpha = \tfrac{1}{2} x^\alpha(T)^\intercal G x^\alpha(T) + \\ \frac{1}{2} \int_{t=0}^{T} [x^\alpha(t)^\intercal Q x^\alpha(t) + u(t)^\intercal R u(t)] dt \quad (5.17)$$

where

$$G = G^\intercal \geq 0, Q = Q^\intercal \geq 0$$

and

$$R = R^\intercal > 0$$

Any terminal set is not assumed to be given as well as any control region, that is,

$$g_l(x) \equiv 0$$

and

$$U = R^r$$

The min–max linear quadratic control problem can be formulated now in the form (5.8):

$$\max_{\alpha \in \mathcal{A}} (h^\alpha) \to \min_{u(\cdot) \in \mathcal{U}_{adm}} \quad (5.18)$$

5.3.2 The Hamiltonian Form and the Parameterization of Robust Optimal Controllers

Following the suggested technique, introduce the Hamiltonian

$$\mathcal{H}^\diamond = \sum_{\alpha \in \mathcal{A}} \left[\psi_\alpha^\intercal (A^\alpha x^\alpha + B^\alpha u + d^\alpha) + \tfrac{1}{2} \psi_{\alpha, n+1} (x^{\alpha\intercal} Q x^\alpha + u^\intercal R u) \right] \quad (5.19)$$

and the adjoint variables $\psi_\alpha(t)$ satisfying

$$\begin{cases} \dot{\psi}_\alpha(t) = -\dfrac{\partial}{\partial x^\alpha} \mathcal{H}^\diamond = -A^{\alpha\intercal}(t) \psi_\alpha(t) - \psi_{\alpha,n+1}(t) Q x^\alpha(t) \\ \dot{\psi}_{\alpha,n+1}(t) = 0 \end{cases} \quad (5.20)$$

as well as the transversality condition

$$\begin{cases} \psi_\alpha(T) = -\mu(\alpha) \operatorname{grad} h^\alpha = \\ -\mu(\alpha) \operatorname{grad} \left[x^\alpha(T)^\mathsf{T} G x^\alpha(T) + x^\alpha_{n+1}(T) \right] = -\mu(\alpha) G x^\alpha(T) \\ \psi_{\alpha,n+1}(T) = -\mu(\alpha) \end{cases} \quad (5.21)$$

Here vector $\psi_\alpha(t)$ is defined as

$$\psi_\alpha(t) := \left(\psi_{\alpha,1}(t), \ldots, \psi_{\alpha,n}(t) \right)^\mathsf{T}$$

The robust optimal control $u(t)$ satisfies (5.15) and leads to

$$\sum_{\alpha \in \mathcal{A}} B^{\alpha\mathsf{T}} \psi_\alpha - \left(\sum_{\alpha \in \mathcal{A}} \mu(\alpha) \right) R^{-1} u(t) = 0 \quad (5.22)$$

Since at least one active index exists it follows that

$$\sum_{\alpha \in \mathcal{A}} \mu(\alpha) > 0$$

Taking into account that if $\mu(\alpha) = 0$, then $\dot{\psi}_\alpha(t) = 0$ and $\psi_\alpha(t) \equiv 0$, the following normalized adjoint variable $\tilde{\psi}_\alpha(t)$ can be introduced

$$\tilde{\psi}_{\alpha,i}(t) = \begin{cases} \psi_{\alpha,i}(t) \mu^{-1}(\alpha) & \text{if } \mu(\alpha) > 0 \\ 0 & \text{if } \mu(\alpha) = 0 \end{cases} \quad (5.23)$$
$$i = 1, \ldots, n+1$$

satisfying

$$\begin{cases} \dot{\tilde{\psi}}_\alpha(t) = -\dfrac{\partial}{\partial x^\alpha} \mathcal{H}^\diamond = -A^{\alpha\mathsf{T}}(t) \tilde{\psi}_\alpha(t) - \tilde{\psi}_{\alpha,n+1}(t) Q x^\alpha(t) \\ \dot{\tilde{\psi}}_{\alpha,n+1}(t) = 0 \end{cases} \quad (5.24)$$

with the transversality conditions given by

$$\begin{cases} \tilde{\psi}_\alpha(T) = -G x^\alpha(T) \\ \tilde{\psi}_{\alpha,n+1}(T) = -1 \end{cases} \quad (5.25)$$

The robust optimal control (5.22) becomes

$$\begin{aligned} u(t) &= \left(\sum_{\alpha \in \mathcal{A}} \mu(\alpha) \right)^{-1} R^{-1} \sum_{\alpha \in \mathcal{A}} \mu(\alpha) B^{\alpha\mathsf{T}} \tilde{\psi}_\alpha \\ &= R^{-1} \sum_{\alpha \in \mathcal{A}} \lambda_\alpha B^{\alpha\mathsf{T}} \tilde{\psi}_\alpha \end{aligned} \quad (5.26)$$

where the vector $\boldsymbol{\lambda} := (\lambda_{\alpha,1}, \ldots, \lambda_{\alpha,N})^\mathsf{T}$ belongs to the simplex S^N defined as

$$S^N := \left\{ \boldsymbol{\lambda} \in R^{N=|\mathcal{A}|} : \lambda_\alpha = \frac{\mu(\alpha)}{\sum_{\alpha=1}^N \mu(\alpha)} \geq 0,\ \sum_{\alpha=1}^N \lambda_\alpha = 1 \right\} \quad (5.27)$$

5.3.3 The Extended Form for the Closed-Loop System

For simplicity, the time argument in the expressions below will be omitted. Introduce the block-diagonal $\mathbb{R}^{nN \times nN}$ valued matrices $\mathbf{A}, \mathbf{Q}, \mathbf{G}, \mathbf{\Lambda}$ and the extended matrix $\mathbf{B}$ as follows:

$$\mathbf{A} := \begin{bmatrix} A^1 & 0 & \cdots & 0 \\ \cdot & \cdot & & \cdot \\ 0 & \cdots & 0 & A^N \end{bmatrix},\ \mathbf{Q} := \begin{bmatrix} Q & 0 & \cdots & 0 \\ \cdot & \cdot & & \cdot \\ 0 & \cdots & 0 & Q \end{bmatrix}$$

$$\mathbf{G} := \begin{bmatrix} G & 0 & \cdots & 0 \\ 0 & \cdot & & 0 \\ 0 & \cdots & 0 & G \end{bmatrix},\ \mathbf{\Lambda} := \begin{bmatrix} \lambda_1 I_{n \times n} & 0 & \cdots & 0 \\ 0 & \cdot & & 0 \\ 0 & & \cdots & 0 & \lambda_N I_{n \times n} \end{bmatrix} \quad (5.28)$$

and

$$\mathbf{B}^\mathsf{T} := \begin{bmatrix} B^{1\mathsf{T}} & \cdots & B^{N\mathsf{T}} \end{bmatrix} \in \mathbb{R}^{r \times nN}$$

In view of these definitions, the dynamic equations (5.16) and (5.24) can be rewritten as

$$\begin{cases} \dot{\mathbf{x}} = \mathbf{A}\mathbf{x} + \mathbf{B}u + \mathbf{d} \\ \mathbf{x}^\mathsf{T}(0) = \left(x^{1\mathsf{T}}(0), \ldots, x^{N\mathsf{T}}(0) \right) \\ \dot{\boldsymbol{\psi}} = -\mathbf{A}^\mathsf{T}\boldsymbol{\psi} + \mathbf{Q}\mathbf{x} \\ \boldsymbol{\psi}(T) = -\mathbf{G}\mathbf{x}(T) \\ u = R^{-1}\mathbf{B}^\mathsf{T}\boldsymbol{\Lambda}\boldsymbol{\psi} \end{cases} \quad (5.29)$$

where

$$\mathbf{x}^\mathsf{T} := \left(x^{1\mathsf{T}}, \ldots, x^{N\mathsf{T}} \right) \in \mathbb{R}^{1 \times nN}$$
$$\boldsymbol{\psi}^\mathsf{T} := \left(\tilde{\psi}_1^\mathsf{T}, \ldots, \tilde{\psi}_N^\mathsf{T} \right) \in \mathbb{R}^{1 \times nN}$$
$$\mathbf{d}^\mathsf{T} := \left(d^{1\mathsf{T}}, \ldots, d^{N\mathsf{T}} \right) \in \mathbb{R}^{1 \times nN}$$

5.3.4 The Robust LQ Optimal Control

Theorem 5.2. *The robust LQ optimal control (5.22) realizing (5.18) is equal to*

$$u = -R^{-1}\mathbf{B}^\mathsf{T}\left[\mathbf{P}_\lambda \mathbf{x} + \mathbf{p}_\lambda\right] \tag{5.30}$$

where the matrix $\mathbf{P}_\lambda = \mathbf{P}_\lambda^T \in \mathbb{R}^{nN \times nN}$ *is the solution of the **parameterized differential matrix Riccati equation***

$$\begin{cases} \dot{\mathbf{P}}_\lambda + \mathbf{P}_\lambda \mathbf{A} + \mathbf{A}^\mathsf{T}\mathbf{P}_\lambda - \mathbf{P}_\lambda \mathbf{B} R^{-1}\mathbf{B}^\mathsf{T}\mathbf{P}_\lambda + \mathbf{\Lambda Q} = 0 \\ \mathbf{P}_\lambda(T) = \mathbf{\Lambda G} = \mathbf{G\Lambda} \end{cases} \tag{5.31}$$

*and the **shifting vector** $\mathbf{p}_\lambda$ satisfies*

$$\begin{cases} \dot{\mathbf{p}}_\lambda + \mathbf{A}^\mathsf{T}\mathbf{p}_\lambda - \mathbf{P}_\lambda \mathbf{B} R^{-1}\mathbf{B}^\mathsf{T}\mathbf{p}_\lambda + \mathbf{P}_\lambda \mathbf{d} = 0 \\ \mathbf{p}_\lambda(T) = 0 \end{cases} \tag{5.32}$$

The matrix $\mathbf{\Lambda} = \mathbf{\Lambda}(\lambda^*)$ *is defined by (5.28) with the weight vector* $\lambda = \lambda^*$ *solving the following finite-dimensional optimization problem:*

$$\lambda^* = \arg\min_{\lambda \in S^N} J(\lambda) \tag{5.33}$$

with

$$\begin{aligned}
J(\lambda) := \max_{\alpha \in \mathcal{A}} h^\alpha &= \frac{1}{2}\left[\mathbf{x}^\mathsf{T}(0)\mathbf{P}_\lambda(0)\mathbf{x}(0) - \mathbf{x}^\mathsf{T}(T)\mathbf{\Lambda G}\mathbf{x}(T)\right] \\
&\quad - \frac{1}{2}\int_0^T \mathbf{x}^\mathsf{T}(t)\mathbf{\Lambda Q}\mathbf{x}(t)dt + \\
&\quad \frac{1}{2}\max_{i=\overline{1,N}}\left[\operatorname{tr}\left\{\left[\int_0^T x^i(t)x^{i\mathsf{T}}(t)dt\right]Q + x^i(T)x^{i\mathsf{T}}(T)G\right\}\right] \\
&\quad + \mathbf{x}^\mathsf{T}(0)\mathbf{p}_\lambda(0) + \frac{1}{2}\int_{t=0}^T \left[2\mathbf{d}^\mathsf{T}\mathbf{p}_\lambda - \mathbf{p}_\lambda^\mathsf{T}\mathbf{B}R^{-1}\mathbf{B}^\mathsf{T}\mathbf{p}_\lambda\right]dt
\end{aligned} \tag{5.34}$$

Proof. Since the robust optimal control (5.29) turns out to be proportional to $\mathbf{\Lambda}\psi$, let us try to find the solution for ψ as follows:

$$\mathbf{\Lambda}\psi = -\mathbf{P}_\lambda \mathbf{x} - \mathbf{p}_\lambda \tag{5.35}$$

The commutation property of the operators

$$\mathbf{\Lambda A}^\mathsf{T} = \mathbf{A}^\mathsf{T}\mathbf{\Lambda}, \quad \mathbf{\Lambda}^k \mathbf{Q} = \mathbf{Q}\mathbf{\Lambda}^k \; (k \geq 0)$$

implies

$$\begin{aligned}
\mathbf{\Lambda}\dot{\psi} &= -\dot{\mathbf{P}}_\lambda \mathbf{x} - \mathbf{P}_\lambda\left[\mathbf{A}\mathbf{x} + \mathbf{B}u + \mathbf{d}\right] - \dot{\mathbf{p}}_\lambda = \\
&\quad -\dot{\mathbf{P}}_\lambda \mathbf{x} - \mathbf{P}_\lambda\left(\mathbf{A}\mathbf{x} - \mathbf{B}R^{-1}\mathbf{B}^\mathsf{T}\left[\mathbf{P}_\lambda\mathbf{x}+\mathbf{p}_\lambda\right] + \mathbf{d}\right) - \dot{\mathbf{p}}_\lambda = \\
&\left[-\dot{\mathbf{P}}_\lambda - \mathbf{P}_\lambda \mathbf{A} + \mathbf{P}_\lambda \mathbf{B}R^{-1}\mathbf{B}^\mathsf{T}\mathbf{P}_\lambda\right]\mathbf{x} + \left(\mathbf{P}_\lambda \mathbf{B}R^{-1}\mathbf{B}^\mathsf{T}\mathbf{p}_\lambda - \mathbf{P}_\lambda \mathbf{d} - \dot{\mathbf{p}}_\lambda\right)
\end{aligned}$$

5.3 Min–Max Linear Quadratic Multimodel Control

$$= [-\Lambda \mathbf{A}^\mathsf{T} \psi + \Lambda \mathbf{Q}\mathbf{x}] = [-\mathbf{A}^\mathsf{T} \Lambda \psi + \Lambda \mathbf{Q}\mathbf{x}]$$
$$= \mathbf{A}^\mathsf{T} [\mathbf{P}_\lambda \mathbf{x} + \mathbf{p}_\lambda] + \Lambda \mathbf{Q}\mathbf{x} = \mathbf{A}^\mathsf{T} \mathbf{P}_\lambda \mathbf{x} + \mathbf{A}^\mathsf{T} \mathbf{p}_\lambda + \Lambda \mathbf{Q}\mathbf{x}$$

or, in the equivalent form,

$$\left[\dot{\mathbf{P}}_\lambda + \mathbf{P}_\lambda \mathbf{A} + \mathbf{A}^\mathsf{T} \mathbf{P}_\lambda - \mathbf{P}_\lambda \mathbf{B} R^{-1} \mathbf{B}^\mathsf{T} \mathbf{P}_\lambda + \Lambda \mathbf{Q} \right] \mathbf{x}$$
$$+ \left[\mathbf{A}^\mathsf{T} \mathbf{p}_\lambda - \mathbf{P}_\lambda \mathbf{B} R^{-1} \mathbf{B}^\mathsf{T} \mathbf{p}_\lambda + \mathbf{P}_\lambda \mathbf{d} + \dot{\mathbf{p}}_\lambda \right] = 0$$

These equations are fulfilled identically under the conditions (5.31) and (5.32) of this theorem. This implies

$$J(\lambda) := \max_{\alpha \in \mathcal{A}} h^\alpha = \max_{\nu \in S^N} \sum_{i=1}^{N} \nu_i h^i =$$

$$\frac{1}{2} \max_{\nu \in S^N} \sum_{i=1}^{N} \nu_i \left[\int_0^T \left[u^\mathsf{T} R u + x^{i\mathsf{T}} Q x^i \right] dt + x^{i\mathsf{T}}(T) G x^i(T) \right] =$$

$$\frac{1}{2} \max_{\nu \in S^N} \int_0^T \left(u^\mathsf{T} R u + \mathbf{x}^\mathsf{T} \mathbf{Q}_\nu \mathbf{x} \right) dt + \mathbf{x}^\mathsf{T}(T) \mathbf{G}_\nu \mathbf{x}(T)$$

where

$$\mathbf{Q}_\nu := \begin{bmatrix} \nu_1 Q & 0 & \cdot & 0 \\ 0 & \cdot & \cdot & \cdot \\ \cdot & \cdot & \cdot & 0 \\ 0 & \cdot & 0 & \nu_N Q \end{bmatrix}, \quad \mathbf{G}_\nu := \begin{bmatrix} \nu_1 G & 0 & \cdot & 0 \\ 0 & \cdot & \cdot & \cdot \\ \cdot & \cdot & \cdot & 0 \\ 0 & \cdot & 0 & \nu_N G \end{bmatrix}$$

and, hence,

$$J(\lambda) = \frac{1}{2} \max_{\nu \in S^N} \left[\int_0^T \left([u^\mathsf{T} \mathbf{B}^\mathsf{T} + \mathbf{x}^\mathsf{T} \mathbf{A} + \mathbf{d}^\mathsf{T}] \Lambda \psi - \mathbf{x}^\mathsf{T} [\mathbf{A} \Lambda \psi - \mathbf{Q}_\nu \mathbf{x}] - \mathbf{d}^\mathsf{T} \Lambda \psi \right) dt \right.$$
$$\left. + \mathbf{x}^\mathsf{T}(T) \mathbf{G}_\nu \mathbf{x}(T) \right] =$$
$$\frac{1}{2} \max_{\nu \in S^N} \left[\int_0^T \left(\dot{\mathbf{x}}^\mathsf{T} \Lambda \psi + \mathbf{x}^\mathsf{T} \Lambda \dot{\psi} + \mathbf{x}^\mathsf{T} \mathbf{Q}_{\nu-\lambda} \mathbf{x} - \mathbf{d}^\mathsf{T} \Lambda \psi \right) dt + \mathbf{x}^\mathsf{T}(T) \mathbf{G}_\nu \mathbf{x}(T) \right]$$

$$= \frac{1}{2} \max_{\nu \in S^N} \left[\int_0^T \left(d\left(\mathbf{x}^\mathsf{T} \Lambda \psi \right) + \mathbf{x}^\mathsf{T} \mathbf{Q}_{\nu-\lambda} \mathbf{x} - \mathbf{d}^\mathsf{T} \Lambda \psi \right) dt + \right.$$
$$\left. \mathbf{x}^\mathsf{T}(T) \mathbf{G}_\nu \mathbf{x}(T) \right] = \frac{1}{2} \left(\mathbf{x}^\mathsf{T}(T) \Lambda \psi(T) - \mathbf{x}^\mathsf{T}(0) \Lambda \psi(0) \right) -$$
$$\frac{1}{2} \int_0^T \left(\mathbf{x}^\mathsf{T} \mathbf{Q}_\lambda \mathbf{x} - \mathbf{d}^\mathsf{T} \left(P_\lambda \mathbf{x} + \mathbf{p}_\lambda \right) \right) dt +$$
$$\frac{1}{2} \max_{\nu \in S^N} \left[\int_0^T \mathbf{x}^\mathsf{T} \mathbf{Q}_\nu \mathbf{x} \, dt + \mathbf{x}^\mathsf{T}(T) \mathbf{G}_\nu \mathbf{x}(T) \right]$$

Thus we obtain
$$J(\lambda) = \frac{1}{2}\left(\mathbf{x}^\mathsf{T}(0)P_\lambda(0)\mathbf{x}(0) - \mathbf{x}^\mathsf{T}(T)G_\lambda\mathbf{x}(T) + \mathbf{x}^\mathsf{T}(0)\mathbf{p}(0)\right)$$
$$- \frac{1}{2}\int_0^T \left(\mathbf{x}^\mathsf{T}\mathbf{Q}_\lambda\mathbf{x} - \mathbf{d}^\mathsf{T}(P_\lambda\mathbf{x} + \mathbf{p}_\lambda)\right) dt + \quad (5.36)$$
$$\frac{1}{2}\max_{\nu \in S^N}\left[\int_0^T \mathbf{x}^\mathsf{T}\mathbf{Q}_\nu\mathbf{x}\,dt + \mathbf{x}^\mathsf{T}(T)\mathbf{G}_\nu\mathbf{x}(T)\right]$$

In view of the identity
$$-\mathbf{x}^\mathsf{T}(0)\mathbf{p}_\lambda(0) = \mathbf{x}^\mathsf{T}(T)\mathbf{p}_\lambda(T) - \mathbf{x}^\mathsf{T}(0)\mathbf{p}_\lambda(0) = \int_{t=0}^T d(\mathbf{x}^\mathsf{T}\mathbf{p}_\lambda)$$
$$= \int_{t=0}^T \left[\mathbf{p}_\lambda^\mathsf{T}\left[\mathbf{A}\mathbf{x} - \mathbf{B}R^{-1}\mathbf{B}^\mathsf{T}\left[P\mathbf{x} + \mathbf{p}_\lambda\right] + \mathbf{d}\right] + \mathbf{x}^\mathsf{T}\dot{\mathbf{p}}_\lambda\right] dt$$
$$= \int_{t=0}^T \left[\mathbf{x}^\mathsf{T}\left(\mathbf{A}^\mathsf{T}\mathbf{p}_\lambda + \dot{\mathbf{p}}_\lambda - PBR^{-1}\mathbf{B}^\mathsf{T}\mathbf{p}_\lambda\right) - \mathbf{p}_\lambda^\mathsf{T} BR^{-1}\mathbf{B}^\mathsf{T}\mathbf{p}_\lambda + \mathbf{d}^\mathsf{T}\mathbf{p}_\lambda\right] dt$$
$$= \int_{t=0}^T \left[-\mathbf{x}^\mathsf{T} P\mathbf{d} - \mathbf{p}_\lambda^\mathsf{T} BR^{-1}\mathbf{B}^\mathsf{T}\mathbf{p}_\lambda + \mathbf{d}^\mathsf{T}\mathbf{p}_\lambda\right] dt$$

it follows that
$$J(\lambda) = \frac{1}{2}\left[\mathbf{x}^\mathsf{T}(0)P_\lambda(0)\mathbf{x}(0) - \mathbf{x}^\mathsf{T}(T)G_\lambda\mathbf{x}(T)\right] + \mathbf{x}^\mathsf{T}(0)\mathbf{p}_\lambda(0)$$
$$-\frac{1}{2}\int_0^T \mathbf{x}^\mathsf{T}\mathbf{Q}_\lambda\mathbf{x}\,dt + \frac{1}{2}\max_{\nu \in S^N}\left[\int_0^T \mathbf{x}^\mathsf{T}\mathbf{Q}_\nu\mathbf{x}\,dt + \mathbf{x}^\mathsf{T}(T)\mathbf{G}_\nu\mathbf{x}(T)\right]$$
$$+\frac{1}{2}\int_{t=0}^T \left[2\mathbf{d}^\mathsf{T}\mathbf{p}_\lambda - \mathbf{p}_\lambda^\mathsf{T} BR^{-1}\mathbf{B}^\mathsf{T}\mathbf{p}_\lambda\right] dt$$

and the relation (5.36) becomes (5.33). □

5.3.5 Robust Optimal Control for Linear Stationary Systems with Infinite Horizon

Let us consider the class of linear stationary controllable systems (5.16) without exogenous inputs, that is,
$$A^\alpha(t) = A^\alpha,\ B^\alpha(t) = B^\alpha,\ d(t) = 0$$

Then, from (5.32) and (5.34), it follows that $\mathbf{p}_\lambda(t) \equiv 0$ and
$$J(\lambda) := \max_{\alpha \in \mathcal{A}} h^\alpha = \frac{1}{2}\left[\mathbf{x}^\mathsf{T}(0)\mathbf{P}_\lambda(0)\mathbf{x}(0) - \mathbf{x}^\mathsf{T}(T)\mathbf{\Lambda G}\mathbf{x}(T)\right]$$
$$-\frac{1}{2}\int_0^T \mathbf{x}^\mathsf{T}(t)\mathbf{\Lambda Q}\mathbf{x}(t)\,dt + \quad (5.37)$$
$$\frac{1}{2}\max_{i=\overline{1,N}}\left[\mathrm{tr}\left\{\left[\int_0^T x^i(t)x^{i\mathsf{T}}(t)\,dt\right]Q + x^i(T)x^{i\mathsf{T}}(T)G\right\}\right]$$

Hence, if the algebraic Riccati equation

$$\mathbf{P}_\lambda \mathbf{A} + \mathbf{A}^\mathsf{T} \mathbf{P}_\lambda - \mathbf{P}_\lambda \mathbf{B} R^{-1} \mathbf{B}^\mathsf{T} \mathbf{P}_\lambda + \mathbf{\Lambda Q} = 0 \qquad (5.38)$$

has a positive definite solution $\mathbf{P}_\lambda$ (the pair $(\mathbf{A}, R^{-1/2}\mathbf{B}^\mathsf{T})$ should be controllable; the pair $(\mathbf{\Lambda}^{1/2}\mathbf{Q}^{1/2}, \mathbf{A})$ should be observable; see, e.g., [52]) for any λ from some subset $S_0^N \subseteq S^N$, then the corresponding closed-loop systems turn out to be stable ($x^\alpha(t) \underset{t\to\infty}{\to} 0$) and the integrals in the right-hand side of (5.37) converge, i.e.,

$$\begin{aligned} J(\lambda) := \max_{\alpha \in \mathcal{A}} h^\alpha &= \frac{1}{2}\left[\mathbf{x}^\mathsf{T}(0)\mathbf{P}_\lambda(0)\mathbf{x}(0) - \mathbf{x}^\mathsf{T}(T)\mathbf{\Lambda G x}(T)\right] \\ &\quad - \frac{1}{2}\int_0^\infty \mathbf{x}^\mathsf{T}(t)\mathbf{\Lambda Q x}(t)dt + \\ &\quad \frac{1}{2}\max_{i=\overline{1,N}}\left[\operatorname{tr}\left\{\left[\int_0^\infty x^i(t)x^{i\mathsf{T}}(t)dt\right]Q + x^i(T)x^{i\mathsf{T}}(T)G\right\}\right] \end{aligned} \qquad (5.39)$$

Corollary 5.1. *The min–max control problem, formulated for the class of multilinear stationary models without exogenous inputs and with the quadratic performance index (5.6) within the infinite horizon, in the case when the algebraic Riccati equation has a positive solution $\mathbf{P}_\lambda$ for any $\lambda \in S_0^N \subseteq S^N$, is solved by the following robust optimal control:*

$$u = -R^{-1}\mathbf{B}^\mathsf{T}\mathbf{P}_\lambda \mathbf{x} \qquad (5.40)$$

where $\mathbf{\Lambda}(\lambda^*)$ is defined by (5.28)) with $\lambda^* \in S_0^N \subseteq S^N$ minimizing (5.39).

5.4 Conclusions

- In this chapter the robust maximum principle is applied to the min–max Bolza multimodel problem given in the general form where the cost function contains a terminal term as well as an integral one and furthermore a fixed horizon and a terminal set are considered.
- For the class of stationary models without any external inputs the robust optimal controller is also designed for the infinite horizon problem.
- The necessary conditions for robust optimality are derived for the class of uncertain systems given by an ordinary differential equation with parameters from a given finite set.
- As an illustration of the suggested approach, the min–max linear quadratic multimodel control problem is considered in the details.
- It is shown that the design of the min–max optimal controller is reduced to a finite-dimensional optimization problem given at the corresponding simplex set containing the weight parameters to be found.

6
Multimodel and ISM Control

Abstract Here, an original linear time-varying system with matched and unmatched disturbances and uncertainties is replaced by a finite set of dynamic models such that each one describes a particular uncertain case including exact realizations of possible dynamic equations also as external unmatched bounded disturbances. Such a trade-off between an original uncertain linear time-varying dynamic system and a corresponding higher-order multimodel system containing only matched uncertainties leads to a linear multimodel system with known unmatched bounded disturbances and unknown matched disturbances as well. Each model from a given finite set is characterized by a quadratic performance index. The developed min–max integral sliding mode control strategy gives an optimal min–max linear quadratic (LQ) control with additional integral sliding mode term. The design of this controller is reduced to a solution of an equivalent min–max LQ problem that corresponds to the weighted performance indices with weights from a finite-dimensional simplex. The additional integral sliding mode controller part completely dismisses the influence of matched uncertainties from the initial time instant. Two numerical examples illustrate this study.

6.1 Motivation

The purpose of this chapter is to take advantage of both techniques used in previous chapters: the min–max robust optimal control and the ISM control. As we have seen, optimal control requires the knowledge of the dynamic equations, here is where the ISM control plays an important role since using it allows to implement the optimal control without affecting the nominal performance of the system. Here we will consider an uncertain system in two senses: we only know that the parameters of the system belong to a finite set and that matched disturbances affect the system.

6.2 Problem Formulation

Let us consider a controlled linear uncertain system

$$\dot{x}(t) = A(t)x(t) + B(t)u(t) + \zeta(t), \quad x(0) = x_0 \qquad (6.1)$$

where $x(t) \in \mathbb{R}^n$ is the state vector at time $t \in [0, T]$, $u(t) \in \mathbb{R}^m$ is a control action, and ζ is an external disturbance (or uncertainty). We will assume that

A6.1 Matrix $B(t)$ is known, it has full rank for all $t \geq 0$ and its pseudoinverse matrix B^+ is bounded:

$$\operatorname{rank} B(t) = m, \quad \left\|[B(t)]^+\right\| < b^+, \quad [B(t)]^+ := [B^\mathsf{T}(t)B(t)]^{-1} B^\mathsf{T}(t)$$

Matrix $A(t)$ may take on the value of a matrix function in a finite number of fixed and a priory known matrix functions, i.e.,

$$A(t) \in \{A^1(t), A^2(t), \ldots, A^N(t)\}$$

where N is a finite number of *possible dynamic scenarios*, here $A^\alpha(t)$ ($\alpha = \overline{1, N}$) is supposed to be bounded, that is,

$$\sup_{t \geq 0} \sup_{\alpha = \overline{1,N}} \|A^\alpha(t)\| < a^+ \qquad (6.2)$$

A6.2 The external disturbances ζ are represented in the following manner:

$$\zeta(t) = \phi(t, x) + \xi(t), \quad t \in [0, T] \qquad (6.3)$$

where $\phi(\cdot)$ is an unmeasured smooth uncertainty, representing the perturbations, which satisfies the *matching condition*, i.e., there exists $\gamma(x, t)$ such that

$$\phi(x, t) = B\gamma(x, t)$$

and $\gamma(x, t)$ is assumed to be bounded as

$$\|\gamma(x, t)\| \leq q\|x\| + p, \qquad q, p > 0 \qquad (6.4)$$

and $\xi(t)$ is an uncertainty taking the finite number of alternative functions, that is, $\xi(t) \in \Xi =: \{\xi^1(t), \ldots, \xi^N(t)\}$ where $\xi^\alpha(t)$ ($\alpha = \overline{1, N}$) are known (smooth enough) bounded functions such that $\|\xi(t)\| \leq \xi^+$ for all $t \in [0, T]$.

Thus, for each concrete realization of possible scenarios, we obtain the following dynamics:

$$\dot{x}^\alpha(t) = A^\alpha(t)x^\alpha(t) + B(t)u(t) + \phi(x^\alpha, t) + \xi^\alpha(t), \quad x^\alpha(0) = x_0 \qquad (6.5)$$

which resembles (5.16), except for the disturbance ϕ. That is why instead of directly applying the min–max optimal control first we will compensate the matched uncertainties.

The control design problem can be formulated as follows: *design the control $u(t)$ in the form*

$$\left.\begin{aligned} u(t) &= u_0(t) + u_1(t) \\ u_1(t) &= u_{1corr} + u_{1comp} \end{aligned}\right\} \quad (6.6)$$

Control $u_1(x,t)$ is a term named the ISM control part. u_{1comp} is responsible for the exact compensation of the unmeasured matched part of $\phi(x,t)$ and $\xi(t)$ from the very beginning of the process. u_{1corr} is a correction term for the linear part of the ISM equations. Control $u_0(x,t)$ is intended to minimize the worst possible scenario in the sense of an LQ index over a finite horizon $t_f \geq 0$, that is,

$$u_0^* = \min_{u_0 \in \mathbb{R}^m} \max_{\alpha = \overline{1,N}} h^\alpha \quad (6.7)$$

$$\begin{aligned} h^\alpha &:= \frac{1}{2}(x^\alpha(t_f))^T L x^\alpha(t_f) + \frac{1}{2}\int_0^{t_f}\Big[(x^\alpha(t))^T Q x^\alpha(t) + \\ &\quad + (u_0(t) + u_{1corr}(t))^T R(u_0(t) + u_{1corr}(t))\Big]dt \\ &\quad Q = Q^\top \geq 0, L = L^\top \geq 0, R = R^\top > 0 \end{aligned} \quad (6.8)$$

Since u_{1comp} is particularly designed for the compensation of matched part of $\phi(x,t)$ and $\xi(t)$, then it is not included in the performance (6.8).

6.3 Design Principles

Substitution of the control laws (6.6) and (6.3) into system (6.1) yields

$$\dot{x}(t) = A(t)x(t) + B(t)u_0(t) + B(t)u_1(t) + \phi(x,t) + \xi(t), \; x(0) = x_0 \quad (6.9)$$

Define the auxiliary *sliding* function $s(x,t) \in \mathbb{R}^m$ as

$$s(x,t) = [B(t)]^+ x(t) - \sigma(t) \quad (6.10)$$

where $\sigma(t)$ represents the integral term which will be defined bellow. Then, it follows that

$$\dot{s}(x,t) = [B(t)]^+ [A(t)x + \xi(t)] + \gamma(x,t) + u_1(t) + u_0(t) - \dot{\sigma}(t) \quad (6.11)$$

Select the auxiliary variable σ as the solution to the differential equation

$$\begin{aligned} \dot{\sigma}(x,t) &= [B(t)]^+ [B(t)u_0(t)] + \left(\frac{d}{dt}[B(t)]^+\right)x \\ \sigma((x(0),0)) &= [B(0)]^+ x(0) \end{aligned} \quad (6.12)$$

Since $A(t) \in \{A^1(t), A^2(t), \ldots, A^N(t)\}$, but we do not know which of these matrices is the matrix of our realization, at difference with the design given in Chap. 2, we cannot include matrix $A(t)$ in the integral term of s (in this case in $\dot{\sigma}$). Then the equation for $s(x,t)$ becomes

$$\dot{s}(x,t) = [B(t)]^+ [A(t)x + \xi(t)] + \gamma(x,t) + u_1(t), \quad s(x,0) = 0 \quad (6.13)$$

In order to achieve *sliding mode dynamics*, let us design the relay control with the form

$$u_1(t) = u_1(x,t) = -M(x)\frac{s}{\|s\|}, \quad M(x) = \bar{q}\|x\| + \bar{p} + \rho, \quad \rho > b^+\xi^+ \quad (6.14)$$

with $\bar{p} \geq p$, $\bar{q} \geq q + b^+ a^+$ (a^+ is a positive constant), which implies

$$\dot{s}(x,t) = [B(t)]^+ \left[B(t)\left(\gamma(x,t) - M(x)\frac{s}{\|s\|}\right) + \xi(t)\right] + [B(t)]^+ A(t)x$$

For the Lyapunov function $V(s) = \frac{1}{2}\|s\|^2$, in view of (6.4) and (6.2), it follows that

$$\frac{d}{dt}V(s) = (s,\dot{s}) = \left(s, \left(\gamma(x,t) - M(x)\frac{s}{\|s\|}\right) + [B(t)]^+ (A(t)x + \xi(t))\right) \leq$$
$$\leq -\|s\|\left(M(x) - \|\gamma(x,t)\| - \left\|[B(t)]^+\right\|\xi^+ - \left\|[B(t)]^+\right\| \|A(t)\| \|x\|\right) \leq$$
$$\leq -\|s\|\left[(\bar{q} - q - b^+a^+)\|x\| + (\bar{p} - p) + \rho - b^+\xi^+\right] \leq -\|s\|\left[\rho - b^+\xi^+\right] \leq 0$$

Thus, in view of (6.12), we derive

$$V(s(x,t)) \leq V(s(x,0)) = \frac{1}{2}\|s(x,0)\|^2 = 0$$

which implies, for all $t \geq 0$, the identities

$$s(x,t) = 0, \; \dot{s}(x,t) = 0 \quad (6.15)$$

It means that the integral sliding mode control (6.14) completely compensates the effect of the matched uncertainty ϕ from the beginning of the process. The relations (6.15) and (6.13) lead to the following representations:

$$u_{1eq} = u_{1corr} + u_{1comp}$$
$$u_{1comp} = -\gamma(x,t) - [B(t)]^+ \xi(t) \text{ and } u_{1corr} = -[B(t)]^+ A(t)x$$

Therefore,

$$\dot{x} = \left[I - B(t)[B(t)]^+\right] A(t)x + B(t)u_0(t) + \left[I - B(t)[B(t)]^+\right] \xi(t) \quad (6.16)$$

Remark 6.1. Define ξ_{eq} as

$$\xi_{eq} = \left[I - BB^+\right] \xi$$

It is clear that

$$\xi_{eq} \in \ker B^+$$

which means that vector ξ_{eq} is a projection of vector ξ onto the space $\ker B^+$.

6.4 Optimal Control Design

Returning to the multimodel case when $A(t)$ may take one of the possible scenarios $A^\alpha(t)$ ($\alpha = \overline{1,N}$), one can conclude that the multimodel system dynamics into the ISM take the form

$$\dot{x}^\alpha(t) = \left[I - B(t)[B(t)]^+\right] A^\alpha(t) x(t) + B(t) u_0(t) \\ + \left[I - B(t)[B(t)]^+\right] \xi^\alpha(t) \tag{6.17}$$

and LQ index (6.8) becomes

$$h^\alpha := \frac{1}{2}\left(x^\alpha(t_f), Lx^\alpha(t_f)\right) + \frac{1}{2}\int_0^{t_f}[(x^\alpha(t), Qx^\alpha(t)) + \\ \left[u_0(t) - \left([B(t)]^+ A^\alpha(t) x^\alpha(t)\right), R\left(u_0(t) - [B(t)]^+ A^\alpha(t) x^\alpha(t)\right)\right]]dt \tag{6.18}$$

The next and last step is to apply the min–max LQ control (see Appendix 5) to the plant (6.17) and obtain the control $u_0(t)$ which together with u_1 (6.14) solves the min–max problem for (6.18).

With the extended system

$$\dot{\mathbf{x}}(t) = \mathbf{A}_{eq}(t)\mathbf{x}(t) + \mathbf{B}(t) u_0(\mathbf{x}, t) + \mathbf{d}$$

and according to Chap. 5, this control is as follows:

$$u_0(\mathbf{x}, t) = -R^{-1}\mathbf{B}^T\left[\mathbf{P}_\lambda \mathbf{x} + \mathbf{p}_\lambda\right] + \mathbf{B}^+ \mathbf{A}\mathbf{\Lambda}\mathbf{x} \tag{6.19}$$

Matrix $\mathbf{P}_\lambda = \mathbf{P}_\lambda^T \in \mathbb{R}^{nN \times nN}$ is the solution of the parameterized differential matrix Riccati equation:

$$\begin{cases} \dot{\mathbf{P}}_\lambda + \mathbf{P}_\lambda\left(\mathbf{A}_{eq} + \mathbf{B}\mathbf{B}^+\mathbf{A}\mathbf{\Lambda}\right) + \left(\mathbf{A}_{eq} + \mathbf{B}\mathbf{B}^+\mathbf{A}\mathbf{\Lambda}\right)^T \mathbf{P}_\lambda - \mathbf{P}_\lambda \mathbf{B} R^{-1}\mathbf{B}^T \mathbf{P}_\lambda + \\ + \mathbf{\Lambda}\left(\mathbf{Q}_{eq} - (\mathbf{B}^+\mathbf{A})^T R \mathbf{B}^+\mathbf{A}\mathbf{\Lambda}\right) = 0; \quad \mathbf{P}_\lambda(t_f) = \mathbf{\Lambda}\mathbf{L} \end{cases} \tag{6.20}$$

and the shifting vector $\mathbf{p}_\lambda$ satisfies

$$\begin{cases} \dot{\mathbf{p}}_\lambda + \left(\mathbf{A}_{eq} + \mathbf{B}\mathbf{B}^+\mathbf{A}\mathbf{\Lambda}\right)^T \mathbf{p}_\lambda - \mathbf{P}_\lambda \mathbf{B} R^{-1} \mathbf{B}^T \mathbf{p}_\lambda + \mathbf{P}_\lambda \mathbf{d} = 0 \\ \mathbf{p}_\lambda(t_f) = 0. \end{cases} \tag{6.21}$$

Here

$$\mathbf{A} := \text{diag}\left(A^1, \ldots, A^N\right), \mathbf{A}_{eq} := \text{diag}\left(A_{eq}^1, \ldots, A_{eq}^N\right), A_{eq}^\alpha = [I - BB^+]A^\alpha \\ \mathbf{Q}_{eq} := \text{diag}\left(Q^1, \ldots, Q^N\right) \\ \mathbf{L} := \text{diag}(L, \ldots, L), \mathbf{\Lambda} := \text{diag}(\lambda_1 I_{n \times n}, \ldots, \lambda_N I_{n \times n}) \\ Q^\alpha = Q + \left[[B(t)]^+ A^\alpha(t)\right]^T R [B(t)]^+ A^\alpha(t) \tag{6.22}$$

and
$$\mathbf{B}^\top := \begin{bmatrix} B(t)^{1T} & \cdots & B(t)^{NT} \end{bmatrix} \in \mathbb{R}^{m \times nN}, \quad \mathbf{B}^+ := \begin{bmatrix} [B(t)]^+ & \cdots & [B(t)]^+ \end{bmatrix}$$
$$\mathbf{d}^\top := \begin{bmatrix} \left(\xi_{eq}^1(t)\right)^T & \cdots & \left(\xi_{eq}^N(t)\right)^T \end{bmatrix} \in \mathbb{R}^{1 \times nN}, \quad \xi_{eq}^\alpha = \begin{bmatrix} I - B(t)[B(t)]^+ \end{bmatrix} \xi^\alpha$$

Matrix $\mathbf{\Lambda} = \mathbf{\Lambda}(\lambda^*)$ is defined by (6.22) with the weight vector $\lambda = \lambda^*$ solving the following finite-dimensional optimization problem:

$$\lambda^* = \arg \min_{\lambda \in \mathbb{S}^N} J(\lambda) \tag{6.23}$$

$$J(\lambda) := \max_{\alpha = \overline{1,N}} h^\alpha = \frac{1}{2} \mathbf{x}^T(0) \mathbf{P}_\lambda(0) \mathbf{x}(0) + \mathbf{x}^T(0) \mathbf{p}_\lambda(0) +$$

$$+ \frac{1}{2} \max_{i=\overline{1,N}} \left[\int_0^{t_f} \left[x^{iT}(t) Q^i x^i(t) + 2x^{iT}(t) \left(B^+ A^i\right)^T \left(\mathbf{B}^\top [\mathbf{P}_\lambda \mathbf{x} + \mathbf{p}_\lambda] - R\mathbf{B}^+ \mathbf{A}\mathbf{\Lambda}\mathbf{x}\right) \right] dt \right.$$

$$\left. + x^{iT}(t_f) L x^i(t_f) \right] -$$

$$\frac{1}{2} \sum_{i=1}^N \lambda_i \left[\int_0^{t_f} \left[x^{iT}(t) Q^i x^i(t) + 2x^{iT}(t) \left(B^+ A^i\right)^T \left(\mathbf{B}^\top [\mathbf{P}_\lambda \mathbf{x} + \mathbf{p}_\lambda] - \right.\right.\right.$$

$$\left.\left.\left. R\mathbf{B}^+ \mathbf{A}\mathbf{\Lambda}\mathbf{x}\right) \right] dt + x^{iT}(t_f) L x^i(t_f) \right] + \frac{1}{2} \int_{t=0}^{t_f} \mathbf{p}_\lambda^\top \left[2\mathbf{d} - BR^{-1}\mathbf{B}^\top \mathbf{p}_\lambda \right] dt$$

$$\mathbb{S}^N = \left\{ \lambda \in \mathbb{R}^N : \lambda_\alpha \geq 0, \sum_{\alpha=1}^N \lambda_\alpha = 1 \right\}$$

A numerical algorithm to find λ^* can be found in Appendix B.2. This means that u_0 is a linear combination of a feedback part (proportional to x) and a shifting vector $\mathbf{p}_\lambda$ which is indeed an open-loop control part. We can summarize the designed control algorithm as follows:

Step 1. for a fixed control u_0, we construct the nominal systems (6.17) and the corresponding LQ index (6.18);
Step 2. construct the control u_0 using the extended system (6.22);
Step 3. design the ISM law u_1 in the form (6.14), compensating the matched part of the uncertainties from the beginning of the process completely;
Step 4. apply the control $u = u_0 + u_1$ to the closed-loop system (6.1).

6.5 Examples

Example 6.1. Let us consider two possible scenarios ($N = 2$) with

$$A^1 = \begin{bmatrix} -0.2t & 2t \\ -0.3t & -1.5t \end{bmatrix}, \quad A^2 = \begin{bmatrix} -0.25t & 2.3t \\ -0.27t & -1.7t \end{bmatrix}$$
$$B^\top = \begin{bmatrix} 2\ t \end{bmatrix}, \quad g^T = \begin{bmatrix} 1.2\sin(4\pi t) & 0.6t\left(\sin 4\pi t\right) \end{bmatrix} \tag{6.24}$$
$$\left(\xi^1\right)^T = \begin{bmatrix} 0.2\sin(\pi t) & 0.25 \end{bmatrix}, \quad \left(\xi^2\right)^T = \begin{bmatrix} 0.5 & 0.3\sin(\pi t) \end{bmatrix}$$

6.6 Linear Time Invariant Case 65

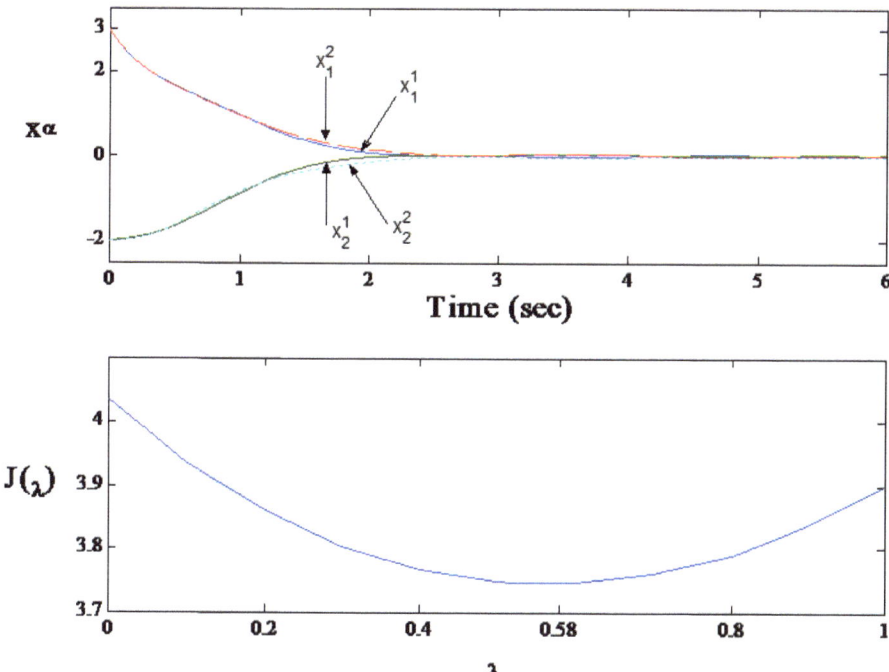

Fig. 6.1. Trajectory of the state variables for system (6.24) and performance index J.

Selecting $R = 1$, $Q = I$, $L = I$, $t_f = 6$, we obtain (see Fig. 6.1) $\lambda_1^* = 0.58$, $\lambda_2^* = 0.42$ and $J(\lambda^*) = 3.744$. The corresponding state variable dynamics are depicted in Fig. 6.1 and the control law is in Fig. 6.2.

Example 6.2. Consider the case of three possible scenarios ($N = 3$) where

$$A^1 = \begin{bmatrix} -1 & 2 \\ 0 & -0.5 \end{bmatrix}, A^2 = \begin{bmatrix} -0.5 & 2.2 \\ 0 & -0.7 \end{bmatrix}, A^3 = \begin{bmatrix} -1.3 & 1.5 \\ 0 & -0.8 \end{bmatrix} \quad (6.25)$$

$$B^T = \begin{bmatrix} 2 & 2 \end{bmatrix}, g^T = \begin{bmatrix} 0.8x_1 & 0.8x_1 \end{bmatrix}, \left(\xi^1\right)^T = \begin{bmatrix} 0.62\sin(2\pi t) & 0.13 \end{bmatrix}$$
$$\left(\xi^2\right)^T = \begin{bmatrix} 0.2 & 0.7 \end{bmatrix}, \left(\xi^3\right)^T = \begin{bmatrix} 0.55 & 0.15 \end{bmatrix}$$

Selecting $R = 1$, $Q = I$, $L = I$, $t_f = 6$ we obtain the optimal weights $\lambda_1^* = 0$, $\lambda_2^* = 0$, $\lambda_3^* = 1$ and the functional $J(\lambda^*) = 4.365$. The corresponding state variable dynamics are shown in Fig. 6.3 and the control law is shown in Fig. 6.4.

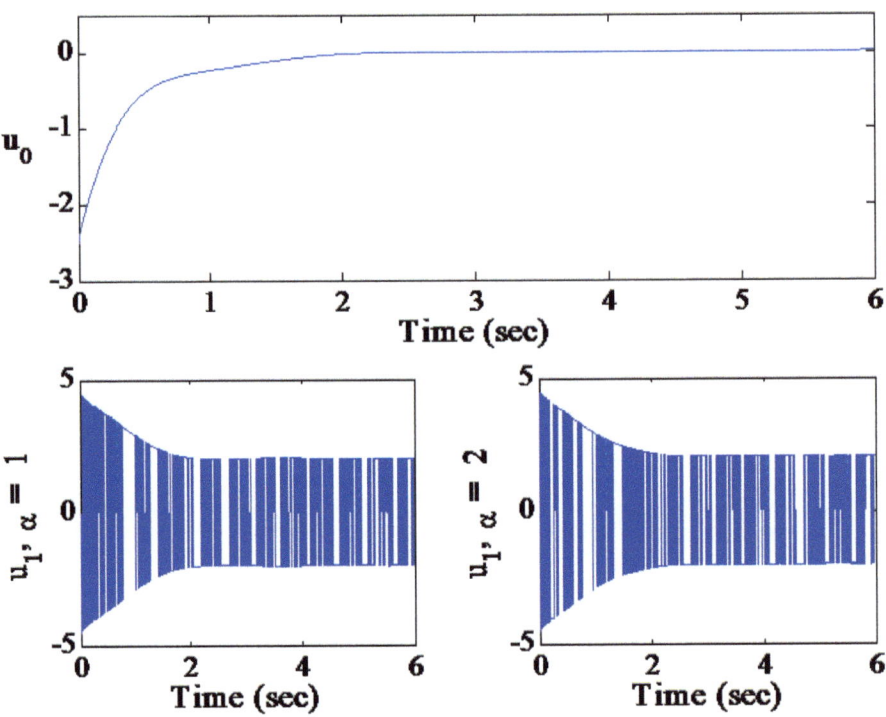

Fig. 6.2. Control u_0 and u_1 for $\alpha = 1$ and $\alpha = 2$.

6.6 Linear Time Invariant Case

The direct usage of ISM in the previous sections requires designing the min–max control law in the space of extended variable with the dimension equal to the product of the state vector's dimension (n) multiplied by the number of scenarios (N), that is, the multimodel optimal problem was solved in the space of nN-order. In this section we design the sliding surface in order to reduce the dimension of the min–max multimodel control design problem (originally equal to $n \cdot N$) up to the space of unmatched uncertainties by $[Nn - (N-1)m]$ dimension (m is the dimension of the control vector).
Let us suppose that system (6.1) is time invariant and that all assumptions are maintained, i.e.,

$$\dot{x}(t) = Ax(t) + Bu_0(t) + Bu_1(x,t) + \phi(x,t) + \xi(t), \; x(0) = x_0 \quad (6.26)$$

Control u_1 is designed following the scheme presented in Sect. 6.3. Thus, system (6.16), in its time invariant version, takes the following form:

$$\dot{x} = A_{eq}x + B(t)u_0(t) + \xi_{eq}(t) \quad (6.27)$$

where $A_{eq} = [I - BB^+]A$ and $\xi_{eq}(t) = [I - BB^+]\xi(t)$.

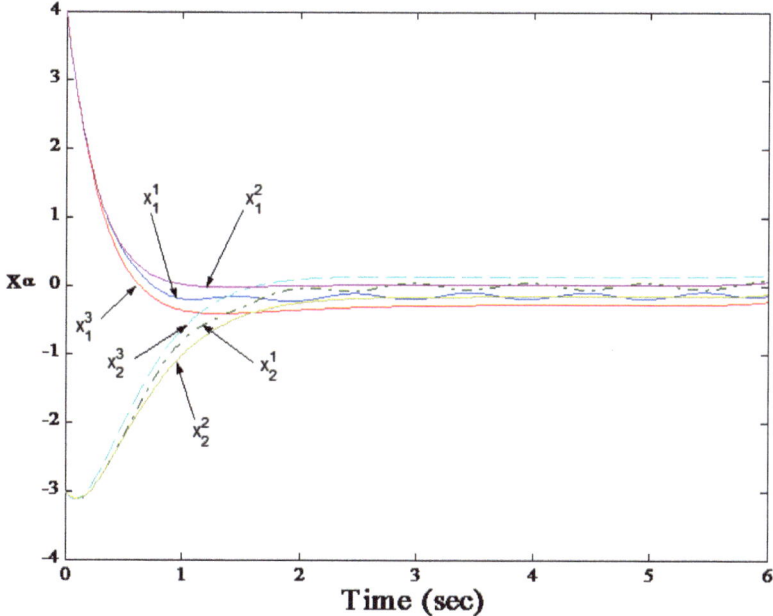

Fig. 6.3. Trajectory of the state variables for system (6.25).

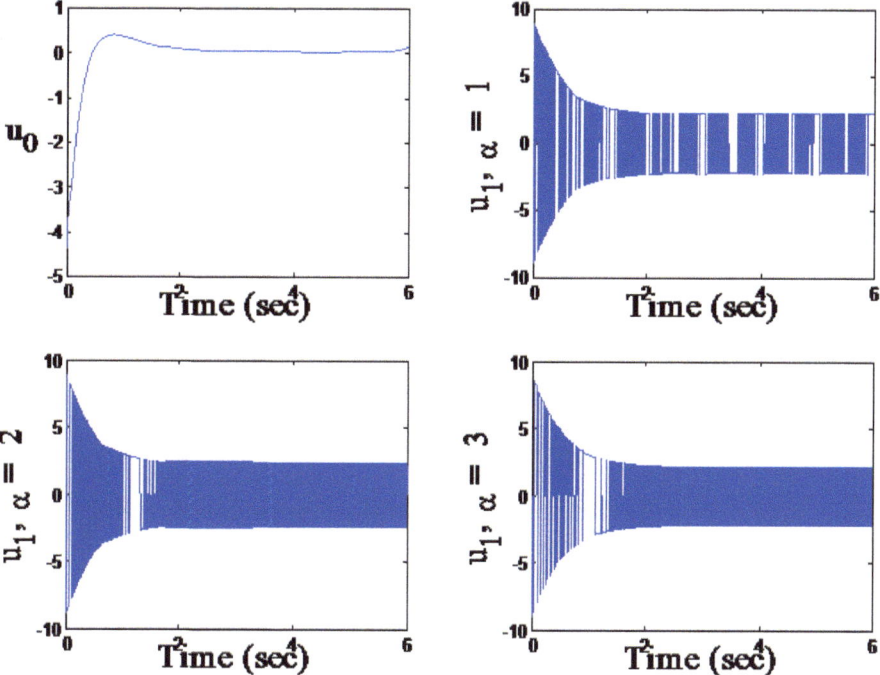

Fig. 6.4. Controls u_0 and u_1 for $\alpha = 1$, $\alpha = 2$, and $\alpha = 3$.

Therefore, the multimodel system dynamics into the ISM is

$$\dot{x}^\alpha(t) = A^\alpha_{eq} x(t) + Bu_0(t) + \xi^\alpha_{eq}(t) \quad (6.28)$$

6.6.1 Transformation of the State

Now, let us transform system (6.27) into two subsystems using the coordinates corresponding to the matched and unmatched parts of uncertainties. Define the following nonsingular transformation:

$$T := \begin{bmatrix} B^\perp \\ B^+ \end{bmatrix}$$

where $B^\perp \in \mathbb{R}^{(n-m) \times n}$ is a matrix which is composed by the transposition of a basis of the orthogonal space of B. Since $\mathrm{rank}(B) = m$, then $\mathrm{rank}(B^\perp) = n - m$.

Applying the transformation T to system (6.27) one obtains

$$z(t) = \begin{bmatrix} z_1(t) \\ z_2(t) \end{bmatrix} := Tx(t) = \begin{bmatrix} B^\perp x(t) \\ B^+ x(t) \end{bmatrix}$$

and

$$\dot{z}(t) = \begin{bmatrix} \dot{z}_1(t) \\ \dot{z}_2(t) \end{bmatrix} = \begin{bmatrix} B^\perp \dot{x}(t) \\ B^+ \dot{x}(t) \end{bmatrix} \quad (6.29)$$

Thus, in the new coordinates, the sliding mode dynamics are governed by the following equations:

$$\dot{z}(t) = \left[T A_{eq} T^{-1} z(t) + T B u_0(t) + T \xi_{eq}(t) \right] =$$

$$\begin{bmatrix} \dot{z}_1(t) \\ \dot{z}_2(t) \end{bmatrix} = \begin{bmatrix} \left[B^\perp A T^{-1} z(t) + B^\perp \xi(t) \right] \\ u_0(t) \end{bmatrix} = \quad (6.30)$$

$$\begin{bmatrix} A_{e1} & A_{e2} \\ 0 & 0 \end{bmatrix} \begin{bmatrix} z_1(t) \\ z_2(t) \end{bmatrix} + \begin{bmatrix} \xi_{e1}(t) \\ 0 \end{bmatrix} + \begin{bmatrix} 0 \\ u_0(t) \end{bmatrix}$$

which may be called *the transformed nominal system*.

6.6.2 The Corrected LQ Index

Let us apply the min–max approach (see the Appendix) to system (6.27), allowing to obtain the control $u_0(x)$ as *a control function minimizing the worst LQ index over a finite horizon t_f*, that is,

$$\min_{u_0 \in \mathbb{R}^m} \max_{\alpha = \overline{1, N}} h^\alpha \quad (6.31)$$

where

$$h^\alpha := \frac{1}{2}(x^\alpha(t_f), Lx^\alpha(t_f)) + \frac{1}{2}\int_{t=0}^{t_f}[(x^\alpha(t), Qx^\alpha(t))$$
$$+ [u_0(t) - (B^+ A^\alpha x^\alpha(t)), R(u(t) - B^+ A^\alpha x^\alpha(t))]]dt$$

$$L = L^T \geq 0,\; Q = Q^T \geq 0,\; R = R^T > 0$$

Since $z(t) = Tx(t)$ and $x(t) = T^{-1}z(t)$, the LQ index h^α can be represented as

$$h^\alpha := \frac{1}{2}(z^\alpha(t_f), (T^T)^{-1} LT^{-1} z^\alpha(t_f)) + \frac{1}{2}\int_{t=0}^{t_f}[(z^\alpha(t), (T^T)^{-1} QT^{-1} z^\alpha(t))+$$
$$+ [u_0(t) - (B^+ A^\alpha T^{-1} z^\alpha(t)), R(u_0(t) - B^+ A^\alpha T^{-1} z^\alpha(t))]]dt$$

6.6.3 Min–Max Multimodel Control Design

Thus, according to [50, 53] (see Appendix at the end of the book), the solution of the optimal problem is as follows. Consider the extended system

$$\dot{\mathbf{x}} = \mathbf{A}_{eq}\mathbf{x} + \mathbf{B}u_0 + \mathbf{d} \tag{6.32}$$

where

$$\mathbf{x}^T = [x^{1T} \cdots x^{NT}],\; \mathbf{A}_{eq} := \mathrm{diag}(A_{eq}^1, \ldots, A_{eq}^N),\; \mathbf{x} \in R^{N \cdot n}$$
$$\mathbf{B}^T := [B^T \cdots B^T],\; \mathbf{d}^T := [\xi_{eq}^{1T} \cdots \xi_{eq}^{NT}] \tag{6.33}$$

Using the previous extended system, the control u_0, denoted below by u_{0_x} to emphasize that it is designed before any state-space transformation application, is as follows:

$$u_{0_x} = -R^{-1}\mathbf{B}^T[\mathbf{P}_\lambda \mathbf{x} + \mathbf{p}_\lambda] + B^+ A\Lambda \mathbf{x} \tag{6.34}$$

where the matrix $\mathbf{P}_\lambda = \mathbf{P}_\lambda^\top \in \mathbb{R}^{nN \times nN}$ is the solution to the parameterized differential matrix Riccati equation:

$$\begin{cases} \dot{\mathbf{P}}_\lambda + \mathbf{P}_\lambda(\mathbf{A}_{eq} + \mathbf{BB}^+ A\Lambda) + (\mathbf{A}_{eq} + \mathbf{BB}^+ A\Lambda)^T \mathbf{P}_\lambda - \mathbf{P}_\lambda \mathbf{B}R^{-1}\mathbf{B}^T\mathbf{P}_\lambda + \\ +\Lambda\left(\mathbf{Q}_{eq} - (B^+ A)^T RB^+ A\Lambda\right) = 0; \qquad \mathbf{P}_\lambda(t_f) = \Lambda L \end{cases} \tag{6.35}$$

and the shifting vector $\mathbf{p}_\lambda$ satisfies

$$\begin{cases} \dot{\mathbf{p}}_\lambda + (\mathbf{A}_{eq} + \mathbf{BB}^+ A\Lambda)^T \mathbf{p}_\lambda - \mathbf{P}_\lambda \mathbf{B}R^{-1}\mathbf{B}^T \mathbf{p}_\lambda + \mathbf{P}_\lambda \mathbf{d} = 0 \\ \mathbf{p}_\lambda(t_f) = 0 \end{cases} \tag{6.36}$$

with the matrices defined as

$$\mathbf{A} := \operatorname{diag}\left(A^1, \ldots, A^N\right), \mathbf{Q}_{eq} := \operatorname{diag}\left(Q^1, \ldots, Q^N\right)$$
$$\mathbf{L} := \operatorname{diag}\left(L, \ldots, L\right), \mathbf{\Lambda} := \operatorname{diag}\left(\lambda_1 I_{n\times n}, \ldots, \lambda_N I_{n\times n}\right)$$
$$Q^\alpha = Q + [B^+ A^\alpha]^T R B^+ A^\alpha.$$

Now consider the extend system using $z(t)$

$$\dot{\mathbf{z}} = \mathbf{T}\mathbf{A}_{eq}\mathbf{T}^{-1}\mathbf{z} + \mathbf{T}\mathbf{B}u_0 + \mathbf{T}\mathbf{d} \qquad (6.37)$$

where

$$z^T = \begin{bmatrix} z^{1T} & \cdots & z^{NT} \end{bmatrix}, \mathbf{T} = \operatorname{diag}\left(T, \ldots, T\right)$$

By (6.37), the control u_0 (denoted by u_{0_z} to emphasize that it is designed after the T-transformation application) is as follows:

$$u_{0_z} = -R^{-1}\left(\mathbf{TB}\right)^T \left[\mathbf{S}_\lambda \mathbf{z} + \mathbf{s}_\lambda\right] + \mathbf{B}^+ \mathbf{A}\mathbf{T}^{-1}\mathbf{\Lambda}\mathbf{z} \qquad (6.38)$$

where the matrix $\mathbf{S}_\lambda = \mathbf{S}_\lambda^T \in \mathbb{R}^{nN \times nN}$ is the solution to the parameterized differential matrix Riccati equation:

$$\begin{cases} \dot{\mathbf{S}}_\lambda + \mathbf{S}_\lambda \left(\mathbf{T}\mathbf{A}_{eq}\mathbf{T}^{-1} + \mathbf{T}\mathbf{B}\mathbf{B}^+\mathbf{A}\mathbf{T}^{-1}\mathbf{\Lambda}\right) + \left(\mathbf{T}\mathbf{A}_{eq}\mathbf{T}^{-1} + \mathbf{T}\mathbf{B}\mathbf{B}^+\mathbf{A}\mathbf{T}^{-1}\mathbf{\Lambda}\right)^T \mathbf{S}_\lambda - \\ -\mathbf{S}_\lambda \mathbf{T}\mathbf{B}R^{-1}\left(\mathbf{TB}\right)^T \mathbf{S}_\lambda + \mathbf{\Lambda}\left(\mathbf{T}^T\right)^{-1}\left(\mathbf{Q}_{eq} - \left(\mathbf{B}^+\mathbf{A}\right)^T R\mathbf{B}^+\mathbf{A}\mathbf{\Lambda}\right)\mathbf{T}^{-1} = 0 \\ \mathbf{S}_\lambda(t_f) = \mathbf{\Lambda}\left(\mathbf{T}^T\right)^{-1}\mathbf{L}\mathbf{T}^{-1} \end{cases}$$

(6.39)

and

$$\begin{cases} \dot{\mathbf{s}}_\lambda + \left(\mathbf{T}\mathbf{A}_{eq}\mathbf{T}^{-1} + \mathbf{T}\mathbf{B}\mathbf{B}^+\mathbf{A}\mathbf{T}^{-1}\mathbf{\Lambda}\right)^T \mathbf{s}_\lambda - \mathbf{S}_\lambda \mathbf{T}\mathbf{B}R^{-1}\left(\mathbf{TB}\right)^T \mathbf{s}_\lambda + \mathbf{S}_\lambda \mathbf{T}\mathbf{d} = 0 \\ \mathbf{s}_\lambda(t_f) = 0 \end{cases}$$

(6.40)

Lemma 6.1. *The controls u_{0_x} in (6.34), designed for system (6.32), and u_{0_z} in (6.38), designed for system (6.37), are identical, that is,*

$$u_{0_z} = u_{0_x} \triangleq u_0 \qquad (6.41)$$

Proof. (6.41) is true if

$$-R^{-1}\mathbf{B}^T\left[\mathbf{P}_\lambda \mathbf{x} + \mathbf{p}_\lambda\right] + \mathbf{B}^+\mathbf{A}\mathbf{\Lambda}\mathbf{x} = -R^{-1}\left(\mathbf{TB}\right)^T \left[\mathbf{S}_\lambda \mathbf{z} + \mathbf{s}_\lambda\right] + \mathbf{B}^+\mathbf{A}\mathbf{T}^{-1}\mathbf{\Lambda}\mathbf{z}$$

Since $\mathbf{T}\mathbf{\Lambda} = \mathbf{\Lambda}\mathbf{T}$ by the triangularity of both multipliers, it implies

$$\mathbf{P}_\lambda = \mathbf{T}^T \mathbf{S}_\lambda \mathbf{T} \text{ and } \mathbf{p}_\lambda = \mathbf{T}^T \mathbf{s}_\lambda \qquad (6.42)$$

and, of course, if (6.42) is true, then the equality (6.41) is satisfied. That is why in order to prove (6.41), it is necessary and sufficient to prove (6.42). Premultiplying (6.39) by $\mathbf{T}^T$ and postmultiplying by $\mathbf{T}$ we obtain

6.6 Linear Time Invariant Case

$$\begin{cases} \mathbf{T^T\dot{S}_\lambda T} + \mathbf{T^T S_\lambda T}\left(\mathbf{A}_{eq}+\mathbf{BB^+A\Lambda}\right) + \left(\mathbf{A}_{eq}+\mathbf{BB^+A\Lambda}\right)^T \mathbf{T^T S_\lambda T} - \\ -\mathbf{T^T S_\lambda T}\mathbf{B}R^{-1}\mathbf{B}^T\mathbf{T^T S_\lambda T} + \mathbf{\Lambda}\left(\mathbf{Q}_{eq} - (\mathbf{B^+A})^T R\mathbf{B^+A\Lambda}\right) = 0 \\ \mathbf{T}^T\mathbf{S}_\lambda\left(t_f\right)\mathbf{T} = \mathbf{\Lambda L} \end{cases}$$

The previous differential Riccati equation is equal to (6.35), taking $\mathbf{P}_\lambda = \mathbf{T}^T\mathbf{S}_\lambda\mathbf{T}$. Now, premultiplying (6.40) by $\mathbf{T}^T$, we obtain

$$\begin{cases} \mathbf{T}^T\dot{\mathbf{s}}_\lambda + \left(\mathbf{A}_{eq}+\mathbf{BB^+A\Lambda}\right)^T \mathbf{T}^T\mathbf{s}_\lambda - \mathbf{T}^T\mathbf{S}_\lambda\mathbf{T}\mathbf{B}R^{-1}\mathbf{B}^T\mathbf{T}^T\mathbf{s}_\lambda + \mathbf{T}^T\mathbf{S}_\lambda\mathbf{T}\mathbf{d} = 0 \\ \mathbf{T}^T\mathbf{s}_\lambda\left(t_f\right) = 0 \end{cases}$$

The previous equation is equal to (6.36) when $\mathbf{p}_\lambda = \mathbf{T}^T\mathbf{s}_\lambda$ and $\mathbf{P}_\lambda = \mathbf{T}^T\mathbf{S}_\lambda\mathbf{T}$. Therefore, (6.41) is proven.

□

Since $z^\alpha(0) = z_0$ and $z_2^\alpha = z_2$, system (6.37), by rearranging the component order, can be represented as follows:

$$\dot{\mathbf{z}}_r = \mathbf{A}_r \mathbf{z}_r + \mathbf{B}_r u_0 + \mathbf{d}_r \tag{6.43}$$

$$\mathbf{z}_r = \begin{bmatrix} z_1^1 \\ \vdots \\ z_1^N \\ z_2 \end{bmatrix}, \quad \mathbf{A}_r := \begin{bmatrix} A_{e1}^1 & 0 & \cdots & 0 & A_{e2}^1 \\ \vdots & \ddots & & \vdots & \vdots \\ 0 & 0 & \cdots & A_{e1}^N & A_{e2}^N \\ 0 & 0 & \cdots & 0 & 0 \end{bmatrix} \tag{6.44}$$

$$\mathbf{B}_r^T = \begin{bmatrix} 0 & \cdots & 0 & I_m \end{bmatrix}, \quad \mathbf{z}_r \in \mathbb{R}^{N(n-m)+m}$$

$$\mathbf{d}_r^T = \begin{bmatrix} \xi_{e1}^{1T} & \cdots & \xi_{e1}^{NT} & 0 \end{bmatrix}$$

We note that in (6.44) we reduce the original (nN) dimension of the state vector up to $Nn-(N-1)m$. Hence we can design the control u_0 using system (6.32) or using system (6.37). It seems to be simpler to deal with the latter from a computational point of view.

According to [50, 53] this control is as follows:

$$u_0 = -R^{-1}\mathbf{B}_r^T\left[\bar{\mathbf{P}}_\lambda \mathbf{z}_r + \bar{\mathbf{p}}_\lambda\right] + \mathbf{F\Lambda}\mathbf{z}_r \tag{6.45}$$

where the matrix $\bar{\mathbf{P}}_\lambda = \bar{\mathbf{P}}_\lambda^T \in \mathbb{R}^{[N(n-m)+m]\times[N(n-m)+m]}$ is the solution to the following parameterized differential matrix Riccati equation

$$\begin{cases} \dot{\bar{\mathbf{P}}}_\lambda + \bar{\mathbf{P}}_\lambda\left(\mathbf{A}_r+\mathbf{B}_r\mathbf{F\Lambda}\right) + \left(\mathbf{A}_r+\mathbf{B}_r\mathbf{F\Lambda}\right)^T \bar{\mathbf{P}}_\lambda - \bar{\mathbf{P}}_\lambda\mathbf{B}_r R^{-1}\mathbf{B}_r^T\bar{\mathbf{P}}_\lambda + \\ + \left(\mathbf{\Lambda Q}_r - \mathbf{\Lambda F}^T R\mathbf{F\Lambda}\right) = 0; \qquad\qquad \bar{\mathbf{P}}_\lambda\left(t_f\right) = \mathbf{\Lambda L} \end{cases} \tag{6.46}$$

and the shifting vector $\bar{\mathbf{p}}_\lambda \in \mathbb{R}^{N(n-m)+m}$ satisfies

$$\begin{cases} \dot{\bar{\mathbf{p}}}_\lambda + (\mathbf{A_r} + \mathbf{B_r F \Lambda})^T \bar{\mathbf{p}}_\lambda - \bar{\mathbf{P}}_\lambda \mathbf{B}_r R^{-1} \mathbf{B}_r^T \bar{\mathbf{p}}_\lambda + \bar{\mathbf{P}}_\lambda d_r = 0 \\ \bar{\mathbf{p}}_\lambda(t_f) = 0 \end{cases} \quad (6.47)$$

with

$$\mathbf{F} = \begin{bmatrix} F_1^1 & \cdots & F_1^N & \lambda_1 F_2^1 + \cdots + \lambda_N F_2^N \end{bmatrix}$$

$$F^\alpha := \begin{bmatrix} F_1^\alpha & F_2^\alpha \end{bmatrix} = B^+ A^\alpha T^{-1}, \; F_2^\alpha \in \mathbb{R}^{m \times m}$$

and using the following partitions

$$Q =: \begin{bmatrix} Q_1 & Q_2 \\ Q_2^T & Q_3 \end{bmatrix}, \; Q^\alpha =: \begin{bmatrix} Q_1^\alpha & Q_2^\alpha \\ (Q_2^\alpha)^T & Q_3^\alpha \end{bmatrix}$$

$$L =: \begin{bmatrix} L_1 & L_2 \\ L_2^T & L_3 \end{bmatrix}$$

$$Q_1, L_1 \in \mathbb{R}^{(n-m) \times (n-m)}, \; Q_3, L_3 \in \mathbb{R}^{m \times m}$$

$$Q_1^\alpha = Q_1 + \left(B_2^{-1} A_{21}^\alpha\right)^T R \left(B_2^{-1} A_{21}^\alpha\right)$$

$$Q_2^\alpha = Q_2 + \left(B_2^{-1} A_{21}^\alpha\right)^T R \left(B_2^{-1} A_{22}^\alpha\right)$$

$$Q_3^\alpha = Q_3 + \left(B_2^{-1} A_{22}^\alpha\right)^T R \left(B_2^{-1} A_{22}^\alpha\right)$$

the following matrices are defined:

$$\mathbf{\Lambda} := \mathrm{diag}\,(\lambda_1 I_{n-m}, \lambda_2 I_{n-m}, \ldots, \lambda_N I_{n-m}, I_m)$$

$$\mathbf{\Lambda Q}_r := \begin{bmatrix} \lambda_1 Q_1^1 & 0 \cdots & 0 & \lambda_1 Q_2^2 \\ \vdots & \ddots & \vdots & \vdots \\ 0 & 0 \cdots & \lambda_N Q_1^N & \lambda_N Q_2^N \\ \lambda_1 (Q_2^1)^T & \cdots & \lambda_N (Q_2^N)^T & \lambda_1 Q_3^1 + \cdots + \lambda_N Q_3^N \end{bmatrix} \quad (6.48)$$

$$\mathbf{\Lambda L} := \begin{bmatrix} \lambda_1 L_1 & 0 \cdots & 0 & \lambda_1 L_2 \\ \vdots & \ddots & \vdots & \vdots \\ 0 & 0 \cdots & \lambda_N L_1 & \lambda_N L_2 \\ \lambda_1 L_2^T & \cdots & \lambda_N L_2^T & L_3 \end{bmatrix}$$

Matrix $\mathbf{\Lambda} = \mathbf{\Lambda}(\lambda^*)$ is defined by (6.48) with the weight vector $\lambda = \lambda^*$ solving the following finite-dimensional optimization problem:

$$\lambda^* = \arg \min_{\lambda \in \mathbb{S}^N} J(\lambda) \quad (6.49)$$

$$J(\lambda) := \max_{\alpha=\overline{1,N}} h^{\alpha} = \frac{1}{2}\mathbf{z}_r^T(0)\bar{\mathbf{P}}_\lambda(0)\mathbf{z}_r(0) + \mathbf{z}_r^T(0)\bar{\mathbf{p}}_\lambda(0) +$$

$$+ \frac{1}{2}\max_{i=\overline{1,N}}\left[\int_0^{t_f} x_0^{iT}(t)Q^i x_0^i(t) + 2x_0^{iT}(t) \times (F^i)^T \left(\mathbf{B}_r^T\left[\bar{\mathbf{P}}_\lambda \mathbf{z}_r + \bar{\mathbf{p}}_\lambda\right] - \right.\right.$$

$$\left.\left. R\mathbf{F}\mathbf{\Lambda}\mathbf{z}_r\right)\right]dt + x_0^{iT}(t_f)Lx_0^i(t_f)\right] - \frac{1}{2}\sum_{i=1}^N \lambda_i \left[\int_0^{t_f} \left[x_0^{iT}(t)Q^i x_0^i(t) + 2x_0^{iT}(t) \times \right.\right.$$

$$\left.\left. \times (F^i)^T \left(\mathbf{B}_r^T\left[\bar{\mathbf{P}}_\lambda \mathbf{z}_r + \bar{\mathbf{p}}_\lambda\right] - R\mathbf{F}\mathbf{\Lambda}\mathbf{z}_r\right)\right]dt + x_0^{iT}(t_f)Lx_0^i(t_f)\right]$$

$$+ \frac{1}{2}\int_{t=0}^{t_f} \bar{\mathbf{p}}_\lambda^T\left[2\mathbf{d}_r - \mathbf{B}_r R^{-1}\mathbf{B}_r^T \bar{\mathbf{p}}_\lambda\right]dt$$

$$\mathbb{S}^N = \left\{\boldsymbol{\lambda} \in \mathfrak{R}^N : \lambda_\alpha \geq 0, \sum_{\alpha=1}^N \lambda_\alpha = 1\right\}$$

where λ^* may be calculated by using the numerical algorithm described in Appendix B.2.

6.7 Example

Let us consider the following system:

$$\dot{x}(t) = A^\alpha x(t) + Bu(t) + \phi(x,t) + \xi^\alpha(t)$$

with three possible scenarios ($N=3$), where

$$A^1 = \begin{bmatrix} -1 & 2 \\ 1.2 & -1.5 \end{bmatrix}, \quad A^2 = \begin{bmatrix} 1 & -2 \\ 1.5 & 1 \end{bmatrix}, \quad A^3 = \begin{bmatrix} 0.5 & 2.5 \\ -1.5 & 1 \end{bmatrix}$$

$$B^T = \begin{bmatrix} 1 & 1 \end{bmatrix}, \quad g^T = \begin{bmatrix} 0.8x_1 & 0.8x_1 \end{bmatrix}, \quad \left(\xi^1\right)^T = \begin{bmatrix} 0.25 & 0.15 \end{bmatrix} \quad (6.50)$$

$$\left(\xi^2\right)^T = \begin{bmatrix} 0.12 & 0.57 \end{bmatrix}, \quad \left(\xi^3\right)^T = \begin{bmatrix} 0.45 & 0.25 \end{bmatrix}$$

Step 1. The nominal system has the following parameters and unmatched uncertainties:

$$\dot{z}(t) = \left[TA_{eq}T^{-1}z(t) + TBu_0(t) + T\xi_{eq}(t)\right]$$

where

$$T := \begin{bmatrix} B^\perp \\ B^+ \end{bmatrix} = \begin{bmatrix} -0.7071 & 0.7071 \\ 0.5 & 0.5 \end{bmatrix}$$

$$TA_{eq}^1 T^{-1} = \begin{bmatrix} -2.85 & -0.9192 \\ 0 & 0 \end{bmatrix}, \quad [T(\xi_{eq}^1)]^T = [-0.0707\ 0]$$

$$TA_{eq}^2 T^{-1} = \begin{bmatrix} 1.25 & 2.4749 \\ 0.0 & 0.0 \end{bmatrix}, \quad [T(\xi_{eq}^2)]^T = [0.3182\ 0]$$

$$TA_{eq}^3 T^{-1} = \begin{bmatrix} 0.25 & -2.4749 \\ 0.0 & 0.0 \end{bmatrix}, \quad [T(\xi_{eq}^3)]^T = [-0.1414\ 0]$$

Step 2. Then, now the objective is to design the control u_0 such that

$$\min_{u_0 \in \mathbb{R}^m} \max_{\alpha = \overline{1,3}} h^\alpha$$

selecting $R = 1$, $Q = I$, $L = I$, $t_f = 10$. The LQ index becomes

$$h^\alpha := \frac{1}{2}(x^\alpha(10), x^\alpha(10)) + \frac{1}{2}\int_{t=0}^{10} [(x^\alpha(t), x^\alpha(t)) + (K^\alpha x^\alpha, K^\alpha x^\alpha) + (u_0(t), u_0(t)) - 2(K^\alpha x^\alpha, u_0(t))]\,dt$$

$$K^1 := [0.1061\ 0.3500]\,x^1,\ K^2 := [-1.2374\ 0.7500]\,x^2$$

$$K^3 = [1.5910\ 1.2500]\,x^3$$

Step 3. The control u_0 is designed using the following extended system:

$$\dot{\mathbf{z}}_r = \mathbf{A}_r \mathbf{z}_r + \mathbf{B}_r u_0(\mathbf{z}_r, t) + \mathbf{d}_r$$

$$\mathbf{z}_r^T = [z_1^1\ z_1^2\ z_1^3\ z_2],\quad \mathbf{B}_r^T = [0\ 0\ 0\ 1]$$

$$\mathbf{A}_r = \begin{bmatrix} -2.85 & 0 & 0 & -0.9192 \\ 0 & 1.25 & 0 & 2.4749 \\ 0 & 0 & 0.25 & -2.4749 \\ 0 & 0 & 0 & 0 \end{bmatrix}$$

$$\mathbf{d}_r^T = [-0.0707\ 0.3182\ -0.1414\ 0]$$

The optimal weights are approximately found as $\lambda_1^* = 0$, $\lambda_2^* = 0.1$, and $\lambda_3^* = 0.9$, and the optimal performance index is $J(\lambda^*) = 594.6517$.

The control u_0 was calculated as in (6.34) and also as in (6.38). In both cases it turned out to be the same. This confirms that the proposed decomposition scheme does not affect the value of $J(\lambda^*)$. In this example the dimension of the extended state vector $\mathbf{z}_r$ of the previous extended system is 4, while the dimension of the state vector $\mathbf{z}$ of system (6.37) is equal to 6.

Step 4. Design the ISM law of control with $M = (2\,\|x\| + 0.5)$ (this is only an option; the choice of M depends on the knowledge of the bound of the matched uncertainty), that implies $u_1 = -(2\,\|x\| + 0.5)\frac{s}{\|s\|}$.

6.7 Example   75

Step 5. Applying the control $u = u_0 + u_1$ to each one within the set of the different given scenarios we obtain the corresponding state variable dynamics and the control law which are depicted in Figs. 6.5 and 6.6.

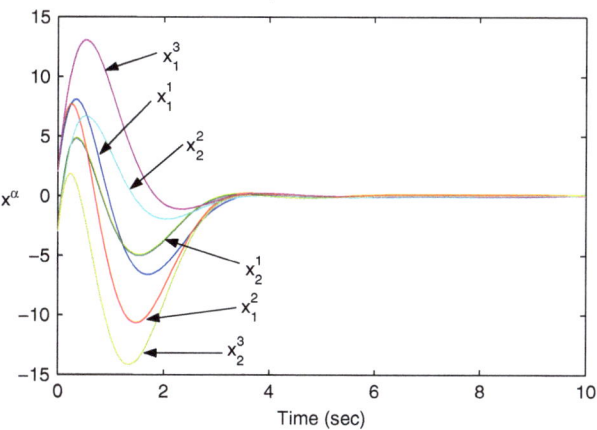

Fig. 6.5. Trajectories of the state variables for the system (6.50).

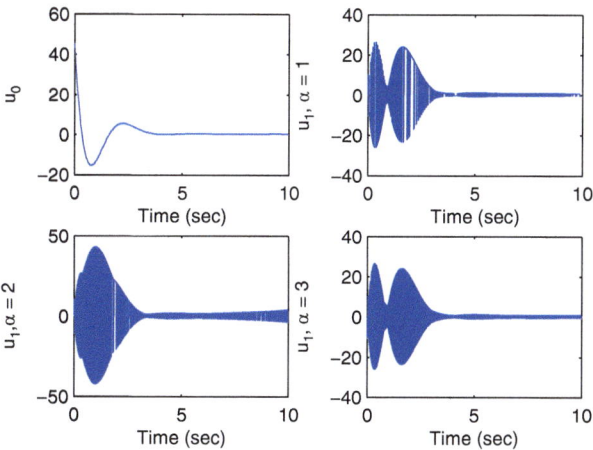

Fig. 6.6. Controls u_0 and u_1.

7
Multiplant and ISM Output Control

Abstract Here, we consider the application of a min–max optimal control based on the LQ index for a set of systems where only the output information is available. Here every system is affected by matched uncertainties, and we propose to use an output integral sliding mode to compensate the matched uncertainties right after the beginning of the process. For the case when the extended system is free of invariant zeros, a hierarchical sliding mode observer is applied. The error of realization of the proposed control algorithm is estimated in terms of the sampling step and actuator time constant. An example illustrates the suggested method of design.

7.1 Motivation

For the multiplant case there are two main approaches to control such systems. One is to decentralize the controls of each plant [54, 55]. The other method is to design the same optimal control law for all the plants and make this control robust with respect to perturbations. As it is explained in the Appendix, in [50, 53], a robust optimal control based on a min–max LQ index for a multimodel system was proposed. Basically, a set of possible models was considered for the same plant, each model is characterized by an LQ index and the objective of the robust optimal control is to minimize the worst of the LQ indexes. However, the exact solution to this optimal control problem requires two basic assumptions:

- the system is free from any uncertainty;
- the state vector is completely available.

Thus, for the case when we have output information only, we should ensure the compensation of the matched uncertainties. Furthermore, we need to reconstruct the original states to take advantage of the state feedback robust optimal control.

The integral sliding mode (ISM) is used to compensate the matched uncertainties from the beginning of the process. However, again, the main problem related to the implementation of this ISM concept consists in the requirement of the *knowledge of the state vector*, including the initial conditions. Thus, the *ISM* turns out to be *useless* if being applied directly in the case when only output information is available.

To realize the robust optimal output control for the multiplant case three approaches must be synthesized:

- the min–max optimal LQ control;
- the integral sliding mode control;
- the hierarchical sliding mode observation.

In Chap. 6 an approach to deal with the problem of matched uncertainty compensation was proposed for the case of a control based on the min–max LQ index in the context of a multimodel system. The difference between multiplant and multimodel systems is the following: in the multimodel case it is considered that for the same plant different models may be realized. However, in the multiplant systems (the case considered in this chapter), we will consider a set of plants working simultaneously and the min–max control law is applied to all plants simultaneously. Moreover in this chapter we will consider that we have no information over the entire state, but only the output of each system can be measured online.

7.2 Problem Formulation

Consider a set of linear time invariant uncertain systems

$$\begin{aligned}\dot{x}^\alpha(t) &= A^\alpha x^\alpha(t) + B^\alpha(u(t) + \gamma(t)) + d^\alpha(t), \quad x^\alpha(0) = x_0^\alpha \\ y^\alpha(t) &= C^\alpha x^\alpha(t)\end{aligned} \quad (7.1)$$

where $\alpha = \overline{1,N}$ (N is a positive integer), $x^\alpha \in \mathbb{R}^n$ represents the state vector for the plant α, $u(t) \in \mathbb{R}^m$ is the vector of control inputs, applied to all the plants, and $y^\alpha(t) \in \mathbb{R}^p$ ($1 \leq p < n$) represents the output vector of each system. Each excitation vector $d^\alpha(t)$ is assumed to be known for all $t \in [0, t_f]$. The current state $x^\alpha(t)$ and the initial state x_0^α are supposed to be non-available. A^α, B^α, and C^α are known matrices of appropriate dimensions with $\operatorname{rank} B^\alpha = m$ and $\operatorname{rank} C^\alpha = p$. All the plants are running in parallel.

Throughout this chapter we will assume that:

A7.1. the vector $\gamma(t)$ is upper bounded by a known scalar function $q_\mathrm{a}(t)$, that is,

$$\|\gamma(t)\| \leq q_\mathrm{a}(t) \quad (7.2)$$

A7.2. every vector x_0^α is bounded by a known constant μ, i.e.,

$$\|x_0^\alpha\| \leq \mu \quad (7.3)$$

Before designing an optimal control we have to make the system free from the effects of matched uncertainties. Therefore, the control design problem can be formulated as follows: *design the control u in the form*

$$u = u_0 + u_1 \tag{7.4}$$

where the control u_1 will compensate the uncertainty $\gamma(t)$ just after the beginning of the process $t = 0$ and $u_0(\cdot)$ is the robust optimal control law $u_0^*(\cdot)$ minimizing the min–max LQ index:

$$\min_{u_0 \in \mathbb{R}^m} \max_{\alpha \in \overline{1,N}} h^\alpha \tag{7.5}$$

$$h^\alpha := \frac{1}{2} \left(x^\alpha(t_f), G^\alpha x^\alpha(t_f) \right) +$$

$$+ \frac{1}{2} \int_{t=0}^{t_f} \left[(x^\alpha(t), Q^\alpha x^\alpha(t)) + (u_0(t), R u_0(t)) \right] dt \tag{7.6}$$

$$Q^\alpha \geq 0,\ G^\alpha \geq 0,\ R > 0$$

along the *nominal* system trajectories

$$\dot{x}^\alpha(t) = A^\alpha x^\alpha(t) + B^\alpha u_0 + d^\alpha \tag{7.7}$$

As an optimal control problem, the exact solution of (7.5) requires the availability of all the vector states $x^\alpha(t)$ at any $t \in [0, t_f]$, and the system must be free of any uncertainty. Therefore, to carry out this optimal control, we firstly should

1. ensure the compensation of the matched uncertainties $\gamma(t)$;
2. design state estimators for each system to reconstruct each state vector $x^\alpha(t)$ practically from the beginning of the process.

7.3 Output Integral Sliding Mode

For each α, substitution of the control law (7.4) into (7.1) yields

$$\dot{x}^\alpha(t) = A^\alpha x^\alpha(t) + B^\alpha (u_0 + u_1 + \gamma(t)) + d^\alpha(t) \tag{7.8}$$

Let us define the following extended system:

$$\begin{aligned} \dot{\mathbf{x}}(t) &= \mathbf{A}\mathbf{x}(t) + \mathbf{B}(u_0 + u_1 + \gamma) + \mathbf{d} \\ \mathbf{y}(t) &= \mathbf{C}\mathbf{x}(t) \end{aligned} \tag{7.9}$$

where

$$\mathbf{x}^T := \begin{bmatrix} x^{1T} & \cdots & x^{NT} \end{bmatrix},\ \mathbf{A} := \mathrm{diag}(A_1, \ldots, A_N),\ \mathbf{B}^T := \begin{bmatrix} B_1^T & \cdots & B_N^T \end{bmatrix}$$
$$\mathbf{C} = \mathrm{diag}(C_1, \ldots, C_N),\ \mathbf{d}^T := \begin{bmatrix} d_1^T & \cdots & d_N^T \end{bmatrix} \tag{7.10}$$

To carry out the OISM, we will also assume that

7 Multiplant and ISM Output Control

A7.3. $\text{rank}(\mathbf{CB}) = m$

Thus, define the auxiliary sliding function s as follows:

$$s(t) := (\mathbf{CB})^+ \mathbf{y}(t) - (\mathbf{CB})^+ \mathbf{y}(0) - \int_0^t \left[(\mathbf{CB})^+ \mathbf{C} [\mathbf{A}\hat{\mathbf{x}}(\tau) + \mathbf{d}(\tau)] - u_0(\tau) \right] d\tau \quad (7.11)$$

where $(\mathbf{CB})^+ = \left[(\mathbf{CB})^T (\mathbf{CB}) \right]^{-1} (\mathbf{CB})^T$. Thus, for the time derivative $\dot{s}$, we have

$$\dot{s} = (\mathbf{CB})^+ \mathbf{CA}(\mathbf{x} - \hat{\mathbf{x}}) + u_0 + u_1 + \gamma, \quad s(0) = 0 \quad (7.12)$$

The vector $\hat{\mathbf{x}}$ represents the state of the observer that will be designed in Sect. 7.4. It is suggested that the control u_1 be designed in the following form:

$$u_1 = -\beta(t) \frac{s(t)}{\|s(t)\|} \quad (7.13)$$

with $\beta(t)$ being a scalar gain satisfying the condition

$$\beta(t) - q_a(t) - \left\| (\mathbf{CB})^+ \mathbf{CA} \right\| \|\mathbf{x} - \hat{\mathbf{x}}\| \geq \lambda_0 > 0$$

where λ is a constant.

Remark 7.1. By A7.2, an upper bound of $\|\mathbf{x} - \hat{\mathbf{x}}\|$ can always be estimated. Indeed, since $\|\mathbf{x} - \hat{\mathbf{x}}\| \leq \|\mathbf{x}\| + \|\hat{\mathbf{x}}\|$, using the Gronwall–Bellman inequality, an upper-bound $\Omega(t, \mathbf{x}(0))$ for $\|\mathbf{x}\|$ can be calculated. Therefore, through the knowledge of $\|\hat{\mathbf{x}}\|$, $\|\mathbf{x} - \hat{\mathbf{x}}\| \leq \Omega(t, \mathbf{x}(0)) + \|\hat{\mathbf{x}}\|$. Nevertheless, this could be a big overestimation, which is why a better way to estimate $\|\mathbf{x} - \hat{\mathbf{x}}\|$ is as follows. The vector $\hat{\mathbf{x}}$ will be given by $\hat{\mathbf{x}} = \tilde{\mathbf{x}} + \mathbf{w}$ where $\tilde{\mathbf{x}}$ represents a Luenberger observer and $\mathbf{w}$ is known and its norm tends to a small constant (see Sects. 7.4.3 and 7.6). Then $\|\mathbf{x} - \tilde{\mathbf{x}}\| < \phi(t) = \zeta \exp(-\kappa t) \left(\sqrt{N}\mu + \|\tilde{\mathbf{x}}(0)\| \right) + \rho$ for positive known constants ζ and κ, and ρ is any arbitrarily small positive constant. Therefore, $\|\mathbf{x} - \hat{\mathbf{x}}\| < \phi(t) + \|\mathbf{w}\|$. Thus, even in the case when $\mathbf{x}$ is unstable, $\|\mathbf{x} - \hat{\mathbf{x}}\|$ has an upper bound which tends to $\rho + \|\mathbf{w}\|$.

As we have done previously, the Lyapunov function $V = \frac{1}{2}\|s\|^2$ is selected to prove the sliding mode existence. Since $\dot{V} = (s, \dot{s})$ and in view of (7.13) and (7.2), one gets

$$\dot{V} = s^T \left((\mathbf{CB})^+ \mathbf{CA}(\mathbf{x} - \hat{\mathbf{x}}) + \gamma - \beta \frac{s}{\|s\|} \right) \leq$$
$$\leq -\|s\| \left(\beta - \left\| (\mathbf{CB})^+ \mathbf{CA} \right\| \|\mathbf{x} - \hat{\mathbf{x}}\| - q_a \right) \leq$$
$$\leq -\|s\| \lambda_0 \leq 0$$

Therefrom, due to $s(0) = 0$, one obtains $\frac{1}{2}\|s(t)\|^2 = V(s(t)) \leq V(s(0)) = \frac{1}{2}\|s(0)\|^2 = 0$. Thus, the identities

$$s(t) = \dot{s}(t) = 0 \tag{7.14}$$

hold for all $t \geq 0$, i.e., there is no reaching phase to the sliding mode.

From (7.12) and in view of the equality (7.14) the *equivalent control* is

$$u_{1eq} = -(\mathbf{CB})^+ \mathbf{CA}(\mathbf{x} - \hat{\mathbf{x}}) - \gamma \tag{7.15}$$

Substitution of u_{1eq} into (7.9) yields

$$\begin{aligned}\dot{\mathbf{x}}(t) &= \tilde{\mathbf{A}}\mathbf{x}(t) + \mathbf{B}(\mathbf{CB})^+ \mathbf{CA}\hat{\mathbf{x}}(t) + \mathbf{B}u_0 + \mathbf{d}(t)\\ \mathbf{y}(t) &= \mathbf{C}\mathbf{x}(t)\end{aligned} \tag{7.16}$$

where

$$\tilde{\mathbf{A}} := \left[I - \mathbf{B}(\mathbf{CB})^+ \mathbf{C}\right]\mathbf{A} \tag{7.17}$$

Thus, our first objective has been achieved, i.e., we have compensated the uncertainty γ. The next section is devoted to the design of the hierarchical sliding mode observer to generate $\hat{\mathbf{x}}$.

7.4 Design of the Observer

Now, having the system without uncertainties, we can reconstruct the state vector. To design the observer, the pair $(\tilde{\mathbf{A}}, \mathbf{C})$ must be observable. The necessary and sufficient condition that guarantees the observability of $(\tilde{\mathbf{A}}, \mathbf{C})$ was given in Lemma 4.2. Therefore, from now on, we will assume that:

A7.4. the triple $(\mathbf{A}, \mathbf{B}, \mathbf{C})$ has no zeros.

It is well known that both conditions together, $\text{rank}(\mathbf{CB}) = m$ and $pN = m$, imply that the triple $(\mathbf{A}, \mathbf{B}, \mathbf{C})$ has zeros. Therefore, A7.3 and A7.4 imply $pN > m$.

As we saw in Chap. 4, the suggested observer is based on the reconstruction of vectors $\mathbf{C}\mathbf{x}(t)$, $\mathbf{C}\tilde{\mathbf{A}}\mathbf{x}(t)$, and so on, until $\mathbf{C}\tilde{\mathbf{A}}^{l-1}\mathbf{x}(t)$. Afterwards, the aim is to recover the vector $\mathbf{Ox}(t)$ where

$$\mathbf{O}^T = \left[\mathbf{C}^T \left(\mathbf{C}\tilde{\mathbf{A}}\right)^T \cdots \left(\mathbf{C}\tilde{\mathbf{A}}^{l-1}\right)^T\right] \tag{7.18}$$

The positive integer l is defined as the observability index, that is, the least positive integer such that $\text{rank}(\mathbf{O}) = n$ (see, e.g., [48]). Thus, after premultiplying $\mathbf{Ox}(t)$ by $\mathbf{O}^+$, the state vector $\mathbf{x}(t)$ can be reconstructed by

82   7 Multiplant and ISM Output Control

$\mathbf{x}(t) = \mathbf{O}^+\mathbf{Ox}(t)$ (vector $\mathbf{Ox}(t)$ is reconstructed online) ($\mathbf{O}^+ = (\mathbf{O}^T\mathbf{O})^{-1}\mathbf{O}^T$ is the pseudoinverse of $\mathbf{O}$).

Design the following dynamic system:

$$\dot{\tilde{\mathbf{x}}}(t) = \tilde{\mathbf{A}}\tilde{\mathbf{x}}(t) + \mathbf{B}u_0(t) + \mathbf{B}(\mathbf{CB})^+\mathbf{CA}\hat{\mathbf{x}}(t) + \\ + \mathbf{L}(\mathbf{y}(t) - \mathbf{C}\tilde{\mathbf{x}}(t)) + \mathbf{d}(t) \quad (7.19)$$

where $\mathbf{L}$ must be designed such that $\hat{\mathbf{A}} := (\tilde{\mathbf{A}} - \mathbf{LC})$ has eigenvalues with negative real part. Let $\mathbf{r}(t) = \mathbf{x}(t) - \tilde{\mathbf{x}}(t)$, from (7.16) and (7.19); the dynamic equations governing $\mathbf{r}(t)$ are

$$\dot{\mathbf{r}}(t) = \left[\tilde{\mathbf{A}} - \mathbf{LC}\right]\mathbf{r}(t) = \hat{\mathbf{A}}\mathbf{r}(t) \quad (7.20)$$

Since the eigenvalues of $\hat{\mathbf{A}}$ have negative real part, (7.20) is exponentially stable, i.e., there exist constants $\gamma,\eta > 0$ such that

$$\|\mathbf{r}(t)\| \leq \gamma e^{-\eta t}\left(\sqrt{N}\mu + \|\tilde{\mathbf{x}}(0)\|\right) \quad (7.21)$$

The Luenberger observer used here ensures the boundedness of the new vector state $\mathbf{r} = \mathbf{x} - \tilde{\mathbf{x}}$. The next step is to reconstruct the error $\mathbf{r}$.

7.4.1 Auxiliary Dynamic Systems and Output Injections

To recover $\mathbf{C}\tilde{\mathbf{A}}\mathbf{x}(t)$, let us introduce an *auxiliary state vector* $\mathbf{x}_a^1(t)$ governed by the following dynamic equations:

$$\dot{\mathbf{x}}_a^1(t) = \tilde{\mathbf{A}}\tilde{\mathbf{x}}(t) + \mathbf{B}\left[u_0(t) + (\mathbf{CB})^+\mathbf{CA}\hat{\mathbf{x}}(t)\right] + \\ + \bar{\mathbf{L}}(\mathbf{C}\bar{\mathbf{L}})^{-1} v^1(t) + \mathbf{d}(t) \quad (7.22)$$

where $\mathbf{x}_a^1(0)$ satisfies $\mathbf{Cx}_a^1(0) = \mathbf{y}(0)$ and $\bar{\mathbf{L}}$ is any matrix such that $\det(\mathbf{C}\bar{\mathbf{L}}) \neq 0$. The vector $\hat{\mathbf{x}}(t)$ represents the observer we will design below. For the variable $s^1(t) \in \mathbb{R}^{Np}$ defined by

$$s^1\left(\mathbf{y}(t), \mathbf{x}_a^1(t)\right) = \mathbf{Cx}(t) - \mathbf{Cx}_a^1(t) \quad (7.23)$$

we have

$$\dot{s}^1\left(\mathbf{y}(t), \mathbf{x}_a^1(t)\right) = \mathbf{C}\tilde{\mathbf{A}}(\mathbf{x}(t) - \tilde{\mathbf{x}}(t)) - v^1(t) \quad (7.24)$$

with $v^1(t)$ defined as $v^1 = M_1 \dfrac{s^1}{\|s^1\|}$. Here the scalar gain M_1 must satisfy the condition $M_1 > \|\mathbf{C}\tilde{\mathbf{A}}\|\|\mathbf{r}\|$ to obtain the sliding mode regime. A bound of $\|\mathbf{r}\|$ can be estimated using (7.21). Hence, with such a scalar gain M_1, we get the identities $s^1(t) = 0$, $\dot{s}^1(t) = 0$, $\forall t \geq 0$. Thus, from (7.23) we obtain that

$$\mathbf{Cx}(t) = \mathbf{C}\tilde{\mathbf{x}}(t), \forall t \geq 0 \quad (7.25)$$

7.4 Design of the Observer

and from (7.24), the equivalent output injection is

$$v_{eq}^1(t) = \mathbf{C}\tilde{\mathbf{A}}\mathbf{x}(t) - \mathbf{C}\tilde{\mathbf{A}}\tilde{\mathbf{x}}(t), \quad \forall t > 0 \tag{7.26}$$

Thus, in principle, $\mathbf{C}\tilde{\mathbf{A}}\mathbf{x}(t)$ can be recovered from (7.26).

Now, the next step is to recover the vector $\mathbf{C}\tilde{\mathbf{A}}^2\mathbf{x}(t)$. Let us design *the second auxiliary state vector* $\mathbf{x}_a^2(t)$ generated by

$$\begin{aligned}\dot{\mathbf{x}}_a^2(t) &= \tilde{\mathbf{A}}^2\tilde{\mathbf{x}}(t) + \tilde{\mathbf{A}}\mathbf{B}u_0(t) + \bar{\mathbf{L}}\left(\mathbf{C}\bar{\mathbf{L}}\right)^{-1}v^2(t) + \\ &+ \tilde{\mathbf{A}}\mathbf{B}(\mathbf{CB})^+\mathbf{C}\mathbf{A}\hat{\mathbf{x}}(t) + \mathbf{d}(t)\end{aligned}$$

where $\mathbf{x}_a^2(0)$ satisfies $\mathbf{C}\tilde{\mathbf{A}}\mathbf{x}_a^1(0) + v_{eq}^1(0) - \mathbf{C}\mathbf{x}_a^2(0) = 0$. Again, for $s^2 \in \mathbb{R}^{Np}$ defined by $s^2\left(v_{eq}^1, \mathbf{x}_a^2\right) = \mathbf{C}\tilde{\mathbf{A}}\tilde{\mathbf{x}}(t) + v_{eq}^1(t) - \mathbf{C}\mathbf{x}_a^2$, and in view of (7.26), we have that s^2 takes the form

$$s^2\left(v_{eq}^1, \mathbf{x}_a^2\right) = \mathbf{C}\tilde{\mathbf{A}}\mathbf{x}(t) - \mathbf{C}\mathbf{x}_a^2 \tag{7.27}$$

Hence, the time derivative of s^2 is

$$\dot{s}^2\left(v_{eq}^1, \mathbf{x}_a^2\right) = \mathbf{C}\tilde{\mathbf{A}}^2\mathbf{x}(t) - \mathbf{C}\tilde{\mathbf{A}}^2\tilde{\mathbf{x}}(t) - v^2(t) \tag{7.28}$$

Now, take the output injection $v^2(t)$ as

$$v^2 = M_2 \frac{s^2}{\|s^2\|}, \quad M_2 > \|\mathbf{C}\tilde{\mathbf{A}}^2\|\|\mathbf{r}\| \tag{7.29}$$

which implies the identities

$$s^2(t) = \dot{s}^2(t) = 0 \tag{7.30}$$

In view of (7.30) and (7.28), $v_{eq}^2(t)$ is

$$v_{eq}^2(t) = \mathbf{C}\tilde{\mathbf{A}}^2\mathbf{x}(t) - \mathbf{C}\tilde{\mathbf{A}}^2\tilde{\mathbf{x}}(t), \quad t > 0 \tag{7.31}$$

and the vector $\mathbf{C}\tilde{\mathbf{A}}^2\mathbf{x}(t)$ can be recovered from (7.31).

By iterating the same procedure, all the vectors $\mathbf{C}\tilde{\mathbf{A}}^i\mathbf{x}$ can be recovered. In a summarized form, the procedure above goes as follows:

(a) *the dynamics of the auxiliary state* $\mathbf{x}_a^k(t)$ *at the kth level are governed by*

$$\begin{aligned}\dot{\mathbf{x}}_a^k(t) &= \tilde{\mathbf{A}}^k\tilde{\mathbf{x}}(t) + \tilde{\mathbf{A}}^{k-1}\mathbf{B}u_0(t) + \bar{\mathbf{L}}\left(\mathbf{C}\bar{\mathbf{L}}\right)^{-1}v^k + \\ &+ \tilde{\mathbf{A}}^{k-1}\mathbf{B}(\mathbf{CB})^+\mathbf{C}\mathbf{A}\hat{\mathbf{x}}(t) + \mathbf{d}(t)\end{aligned} \tag{7.32}$$

with $\bar{\mathbf{L}}$ being any constant matrix such that $\det\left(\mathbf{C}\bar{\mathbf{L}}\right) \neq 0$, and *the output injection* v^k *at the kth level is*

$$v^k = M_k \frac{s^k}{\|s^k\|}, \quad M_k > \|\mathbf{C}\tilde{\mathbf{A}}^k\|\|\mathbf{r}\| \tag{7.33}$$

where M_k is a scalar gain. A bound of $\|\mathbf{r}\|$ may be found using (7.21);

(b) define s^k at the k-level of the hierarchy as

$$s^k(t) = \begin{cases} \mathbf{y} - \mathbf{C}\mathbf{x}_a^1, & k = 1 \\ v_{eq}^{k-1} + \mathbf{C}\tilde{\mathbf{A}}^{k-1}\tilde{\mathbf{x}} - \mathbf{C}\mathbf{x}_a^k, & k > 1 \end{cases} \quad (7.34)$$

where v_{eq}^{k-1} is the *equivalent output injection* whose general expression will be obtained in the following lemma, but $\mathbf{x}_a^k(0)$ should be chosen such that $s^k(0)$ satisfies

$$s^k(0) = 0, \ k = 1, .., l - 1 \quad (7.35)$$

Lemma 7.1. *If the auxiliary state vector $\mathbf{x}_a^k$ and the variable s^k are designed as in (7.32) and (7.34), respectively, then*

$$v_{eq}^k(t) = \mathbf{C}\tilde{\mathbf{A}}^k[\mathbf{x}(t) - \tilde{\mathbf{x}}(t)] \ \text{for all} \ t \geq 0 \quad (7.36)$$

at each $k = \overline{1, l-1}$.

Proof. It was shown that the following identity holds $v_{eq}^1(t) = \mathbf{C}\tilde{\mathbf{A}}[\mathbf{x}(t) - \tilde{\mathbf{x}}(t)]$, $\forall t > 0$. Now, suppose that the equivalent output injection v_{eq}^{k-1} is as (7.36), then the substitution of v_{eq}^{k-1} into (7.34) gives

$$s^k\left(v_{eq}^{k-1}(t), \mathbf{x}_a^k(t)\right) = \mathbf{C}\tilde{\mathbf{A}}^{k-1}\mathbf{x}(t) - \mathbf{C}\mathbf{x}_a^k(t) \quad (7.37)$$

The derivative of (7.37) yields

$$\dot{s}^k(t) = \mathbf{C}\tilde{\mathbf{A}}^k[\mathbf{x}(t) - \tilde{\mathbf{x}}(t)] - v^k(t) \quad (7.38)$$

Thus, selecting $v^k(t)$ as in (7.33), one gets

$$s^k(t) \equiv 0, \ \dot{s}^k(t) \equiv 0 \ \text{for all} \ t \geq 0 \quad (7.39)$$

Therefore, (7.39) and (7.38) imply (7.36). □

7.4.2 Observer in Its Algebraic Form

Now, we can design an observer with the properties required in the problem statement. From (7.25) and (7.36), we obtain the following algebraic equation arrangement:

$$\begin{aligned} \mathbf{C}\mathbf{x}(t) &= \mathbf{C}\tilde{\mathbf{x}}(t) + [\mathbf{y}(t) - \mathbf{C}\tilde{\mathbf{x}}(t)] \\ \mathbf{C}\tilde{\mathbf{A}}\mathbf{x}(t) &= \mathbf{C}\tilde{\mathbf{A}}\tilde{\mathbf{x}}(t) + v_{eq}^1(t) \\ &\vdots \\ \mathbf{C}\tilde{\mathbf{A}}^{l-1}\mathbf{x}(t) &= \mathbf{C}\tilde{\mathbf{A}}^{l-1}\tilde{\mathbf{x}}(t) + v_{eq}^{l-1} \end{aligned} \quad (7.40)$$

Thus, (7.40) yields the matrix equation

$$\mathbf{O}\mathbf{x}(t) = \mathbf{O}\tilde{\mathbf{x}}(t) + v_{eq}(t), \ \forall t > 0 \quad (7.41)$$

where **O** was defined in (7.18) and

$$v_{eq}^T = \left[(\mathbf{y}(t) - \mathbf{C}\tilde{\mathbf{x}}(t))^T \ (v_{eq}^1)^T \ \cdots \ (v_{eq}^{l-1})^T \right] \tag{7.42}$$

Since the pair $\left(\tilde{\mathbf{A}}, \mathbf{C} \right)$ is observable, matrix **O** has rank Nn. Thus, after premultiplying $\mathbf{O}^+$ by (7.41), we obtain

$$\mathbf{x}(t) \equiv \tilde{\mathbf{x}}(t) + \mathbf{O}^+ v_{eq}(t), \quad \forall t > 0 \tag{7.43}$$

Thus, the observer can be designed as

$$\hat{\mathbf{x}}(t) := \tilde{\mathbf{x}}(t) + \mathbf{O}^+ v_{eq}(t) \tag{7.44}$$

Now, we can formulate the following theorem.

Theorem 7.1. *Under the assumptions A7.1–A7.4*

$$\hat{\mathbf{x}}(t) \equiv \mathbf{x}(t) \quad \forall t > 0 \tag{7.45}$$

Proof. It follows directly from (7.43) and (7.44).

□

7.4.3 Observer Realization

As was explained in Chap. 4, to implement the observer described in (7.44), we need to estimate v_{eq}^k, which can be indirectly measured by means of the following first-order low-pass filter:

$$\tau \dot{v}_{av}^k(t) + v_{av}^k(t) = v^k(t); \quad v_{av}^k(0) = 0 \tag{7.46}$$

thereby obtaining an approximation of v_{eq}^k. That is, $\lim_{\substack{\tau \to 0 \\ \delta/\tau \to 0}} v_{av}^k(t) = v_{eq}^k(t)$, $t > 0$, where δ is the sampling time used in the computations during the realization of the observer. So, we can select $\tau = \delta^\eta$ $(0 < \eta < 1)$.

Hence, to realize the HSM observer, we should

(1) use a very small sampling interval δ;
(2) substitute $v_{eq}^k(t)$ into (7.34) and (7.42) by $v_{av}^k(t)$;
(3) choose $\mathbf{x}_a^k(0)$ so that

$$\mathbf{y}(0) - \mathbf{C}\mathbf{x}_a^1(0) = 0, \text{ for } k = 1$$
$$\mathbf{C}\tilde{\mathbf{A}}^{k-1}\tilde{\mathbf{x}}(0) - \mathbf{C}\mathbf{x}_a^k(0) = 0, \text{ for } k > 1$$

Therefore the identity $s^k(0) = 0$, $k = 1, \ldots, l-1$ is achieved.
Thus, with the extended vector formed by the filter outputs, i.e.,

$$\mathbf{v}_{av}^T := \left[(\mathbf{y}(t) - \mathbf{C}\tilde{\mathbf{x}}(t))^T \ (v_{av}^1)^T \ \cdots \ (v_{av}^{l-1})^T \right]$$

the observer $\hat{\mathbf{x}}(t)$ must be redefined as

$$\hat{\mathbf{x}}(t) := \tilde{\mathbf{x}}(t) + \mathbf{H}^+ \mathbf{v}_{av}(t) \tag{7.47}$$

7.5 Min–Max Optimal Control Design

In this section we return back to the problem of the optimal control u_0 which solves the problem (7.5). Substitution of (7.45) into (7.16) yields the sliding motion equations and the state $\mathbf{x}$ takes the form

$$\dot{\mathbf{x}}(t) = \mathbf{A}\mathbf{x}(t) + \mathbf{B}u_0(\mathbf{x}) + \mathbf{d}$$

The control solving (7.5) for (7.7) is of the form

$$u_0^*(\mathbf{x}, t) = -R^{-1}\mathbf{B}^{\mathsf{T}}\left(\mathbf{P}_{\lambda^*}\mathbf{x} + \mathbf{p}_{\lambda^*}\right) \tag{7.48}$$

where the matrix $\mathbf{P}_\lambda \in \mathbb{R}^{nN \times nN}$ is the solution of the parameterized differential matrix Riccati equation:

$$\dot{\mathbf{P}}_\lambda + \mathbf{P}_\lambda \mathbf{A} + \mathbf{A}^T \mathbf{P}_\lambda - \mathbf{P}_\lambda BR^{-1}B^T \mathbf{P}_\lambda + \mathbf{\Lambda Q} = 0 \\ \mathbf{P}_\lambda(t_f) = \mathbf{\Lambda G} \tag{7.49}$$

and the shifting vector $\mathbf{p}_\lambda$ satisfies

$$\dot{\mathbf{p}}_\lambda + \mathbf{A}^{\mathsf{T}}\mathbf{p}_\lambda - \mathbf{P}_\lambda BR^{-1}\mathbf{B}^{\mathsf{T}}\mathbf{p}_\lambda + \mathbf{P}_\lambda \mathbf{d} = 0; \quad \mathbf{p}_\lambda(t_f) = 0 \tag{7.50}$$

where the weighting vector λ belongs to the simplex $\mathbb{S}^N$

$$\mathbb{S}^N := \left\{ \lambda \in \mathbb{R}^N : \lambda_\alpha \geq 0, \sum_{\alpha=1}^{N} \lambda_\alpha = 1 \right\}$$

and the matrices $\mathbf{Q}$, $\mathbf{G}$, and $\mathbf{\Lambda}$ denote the extended matrices

$$\mathbf{Q} := \mathrm{diag}\left(Q_1, \ldots, Q_N\right), \; \mathbf{G} := \mathrm{diag}\left(G_1, \ldots, G_N\right) \\ \mathbf{\Lambda} := \mathrm{diag}\left(\lambda_1 I_n, \ldots, \lambda_N I_n\right) \tag{7.51}$$

The matrix $\mathbf{\Lambda} = \mathbf{\Lambda}(\lambda^*)$ is defined by (7.51) with the weighting vector $\lambda = \lambda^*$ solving the following finite-dimensional optimization problem:

$$\lambda^* = \arg \min_{\lambda \in \mathbb{S}^N} J(\lambda) \\ J(\lambda) := \max_{\alpha = \overline{1,N}} h^\alpha \tag{7.52}$$

From Theorem B.1, the weighting vector λ^* can be generated by means of the sequence $\{\lambda^k\}$ defined by

$$\lambda^{k+1} = \pi \left\{ \lambda^k + \frac{\gamma^k}{J(\lambda^k) + \epsilon} F\left(\lambda^k\right) \right\}, \; \lambda^0 \in \mathbb{S}^N \\ k = 0, 1, 2, \ldots \\ F\left(\lambda^k\right) = \begin{bmatrix} h^1_{\lambda^k} & \cdots & h^N_{\lambda^k} \end{bmatrix}^T, \; J\left(\lambda^k\right) := \max_{\alpha \in \overline{1,N}} h^\alpha_{\lambda^k} \tag{7.53}$$

7.5 Min–Max Optimal Control Design

where ϵ is an arbitrary strictly positive (small enough) constant and $\pi\{\cdot\}$ is the projector to the simplex $\mathbb{S}^N$, i.e., for each $z \in \mathbb{R}^N$,

$$\|\pi\{z\} - z\| < \|\lambda - z\|, \text{ for every } \lambda \in \mathbb{S}^N, \lambda \neq \pi\{z\}$$

From Theorem B.1 we have that

$$\lim_{k \to \infty} \lambda^k = \lambda^* \tag{7.54}$$

provided that the following conditions are satisfied:

(1) for any $\lambda' \neq \lambda'' \in \mathbb{S}^N$ the following inequality holds:

$$\left(\lambda' - \lambda'', F\left(\lambda'\right) - F\left(\lambda''\right)\right) < 0 \tag{7.55}$$

and the identity in (7.55) is possible if and only if $\lambda' = \lambda''$;

(2) there exists a constant L such that for all $\alpha \in \overline{1, N}$ and for any $\mu, \lambda \in \mathbb{S}^N$,

$$|h^\alpha(\mu) - h^\alpha(\lambda)| \leq J(\lambda) L |\mu - \lambda|$$

(3) the gain sequence $\{\gamma^k\}$ satisfies

$$\gamma^k > 0, \ \sum_{k=0}^{\infty} \gamma^k = \infty, \ \sum_{k=0}^{\infty} \left(\gamma^k\right)^2 < \infty$$

Since the observation error can be made arbitrarily small after any arbitrarily small time, the estimated state $\hat{\mathbf{x}}$ can be used instead of $\mathbf{x}$. Therefore, the control u_0 should be designed as

$$u_0\left(\hat{\mathbf{x}}, t\right) = u_0^*\left(\hat{\mathbf{x}}, t\right) = -R^{-1}\mathbf{B}^\mathsf{T}\left[\mathbf{P}_{\lambda^*}\hat{\mathbf{x}} + \mathbf{p}_{\lambda^*}\right] \tag{7.56}$$

with $\hat{\mathbf{x}}$ being designed according to (7.47).

7.5.1 Control Algorithm

The proposed control algorithm can be summarized as follows:

1. design the control u_1 according to (7.13);
2. design the matrix $\mathbf{L}$ such that the eigenvalues of $\hat{\mathbf{A}} := (\tilde{\mathbf{A}} - \mathbf{LC})$ have negative real part;
3. design the auxiliary systems $\mathbf{x}_a^k$ as in (7.32) with the sliding function s^k as in (7.34) and compute the constants M_k, $k = 1, .., l - 1$;
4. design the state estimator $\hat{\mathbf{x}}$ according to (7.47);
5. calculate the matrix $\mathbf{P}_{\lambda^*}$ and the vector $\mathbf{p}_{\lambda^*}$ according to (7.49) and (7.50), respectively;
6. use the sequence (7.53) for finding λ^*, using $\hat{\mathbf{x}}$ instead of $\mathbf{x}$;
7. design u_0 according to (7.56).

7.6 Error Estimation During Implementation of the Closed-Loop Control

We have seen that filters cause some errors in the state estimation. Evidently those errors directly affect the controller since we have used the estimated states instead of the original ones. Hence, here we will calculate the estimation of the error appearing during the realization of the closed-loop control, that is, the errors due to the actuators plus the error due to the observation process. The control error due to the devices used in the implementation of the control, including the OISM control, is of the order $O(\mu)$, where μ is a constant characterizing the control execution depending generally from the actuator time constants. Let us estimate the order of the error due to the observation process. As we saw, the observer design is based on the recursive use of filters of the form (7.46). Firstly, let us recall the following lemma regarding the error induced for such sort of filters.

Lemma 7.2 ([11]). *If in the differential equation*

$$\tau \dot{z} + z = h(t) + H(t)\dot{s} \qquad (7.57)$$

where τ is a constant and z, h, and s are m-dimensional vector functions

(1) the functions $h(t)$ and $H(t)$ and their first-order derivatives are bounded in magnitude by a certain number M and
(2) $\|s(t)\| \leq \xi$ (ξ is a constant positive value)

then for any pair of positive numbers Δt and v there exists a number $d(v, \Delta t, z(0))$ such that

$$\|z(t) - h(t)\| \leq v$$

with $0 < \tau \leq d$, $\xi/\tau \leq d$ and $t \geq \Delta t$.

Indeed, $\|z(t) - h(t)\|$ satisfies the following inequality:

$$\|z(t) - h(t)\| \leq \|z(0) - h(0)\| \exp(-t/\tau) +$$
$$+ M(\tau + \xi) + 3M\left(\frac{\xi}{\tau}\right)$$

In our case, the expression (7.57) can be related with the expression

$$\tau_1 \dot{v}^1_{av} + v^1_{av} = v^1_{eq} - \dot{s}^1$$

obtained from (7.46) and (7.24). Thus, in our case $h(t)$ refers to the equivalent output injection. Furthermore, the error due to sliding mode control directly affects the performance of the first sliding mode in the observation process, and it is known that the sampling step δ induces an error of the order $O(\delta)$ in the variable $\dot{s}^1$ during the sliding motion. Hence, it is reasonable to accept that the error in the sliding variable $\dot{s}^1$ is of the order $O(\mu) + O(\delta)$, that is, by defining $\Delta := \mu + \delta$, we have that the constant ξ in Lemma 7.2 is $\xi = O(\Delta) = O(\mu) + O(\delta)$. Therefore, choosing $\tau = O(\Delta^{1/2})$, we have that the error in the first step of the observation scheme is of the order $O(\Delta^{1/2})$, that is, $v_{av}^1 - v_{eq}^1 = O(\Delta^{1/2})$. As it was mentioned in Sect. 7.4.3, we must substitute v_{eq}^1 by v_{av}^1 into the variable s^2 in (7.27). Thus, we can consider that during the sliding motion, s^2 will be bounded for a constant of order $O(\Delta^{1/2})$, and consequently, by using a constant of the filter $\tau_2 = O(\Delta^{1/4})$, the error induced for the second filter will be $v_{av}^2 - v_{eq}^2 = O(\Delta^{1/4})$. Following a likewise analysis, we obtain an error of the order $O(\Delta^{1/2^k})$ in the kth step for the reconstruction of the observer. Thus, it turns out to be that the observation error is of the order $O(\Delta^{1/2^l})$, recalling that l is the least integer such that matrix $\mathbf{O}$ in (7.18) has rank nN. Thus, we can say that, during the realization of the control process, the total error ϵ_c of the closed-loop control is

$$\epsilon_c = O(\mu) + O\left(\Delta^{1/2^l}\right)$$

7.7 Example

Consider the case $N = 3$ where the parameters are given by

$$A_1 = \begin{bmatrix} -2 & 0.5 & 1 \\ 0.5 & 1.2 & -2 \\ 1 & 2 & -1.5 \end{bmatrix}, A_2 = \begin{bmatrix} -0.3 & 1.5 & -0.15 \\ -1 & 0.12 & 2 \\ 1 & 2 & -3 \end{bmatrix}$$

$$A_3 = \begin{bmatrix} 0.4 & -1 & 0.3 \\ 0.5 & -0.4 & 0.3 \\ 0.5 & 0.6 & -1 \end{bmatrix}, B_1 = \begin{bmatrix} 0.5 \\ 1 \\ 1 \end{bmatrix}, B_2 = \begin{bmatrix} 1.5 \\ -2 \\ 1 \end{bmatrix}$$

$$B_3 = \begin{bmatrix} 0.5 \\ 0.2 \\ 1 \end{bmatrix}, C^\alpha = \begin{bmatrix} 1 & 0 & 0 \\ 0 & 1 & 0 \end{bmatrix}, d^\alpha = \begin{bmatrix} 0.05 \\ 0.02 \\ 0.01 \end{bmatrix}$$

In the simulations we used $\gamma(t) = \sin(t)$ and a sampling time $\delta = 10^{-4}$. Table 7.1 shows the components of the vector λ^k, $k = 1, \ldots, 35$, calculated using the sequence (7.53); it also shows the performance indexes of each plant $h_{\lambda^k}^\alpha$ and the index[1] $J(\lambda^k)$. The trajectories for the three plants are shown

---

[1] It is known that P_λ and p_λ must be precomputed backward using a numerical method. However, the difference here with the classical optimal control is that these equations are parameterized by a weighting vector λ. The calculation of the optimal weighting vector λ^* is not standard. That is why in the example we present a table with the values obtained using the numerical algorithm described above for the calculation of λ^*.

Table 7.1. Values of λ^k and $h^\alpha(\lambda^k)$

k	λ_1^k	λ_2^k	λ_3^k	$h_{\lambda^k}^1$	$h_{\lambda^k}^2$	$h_{\lambda^k}^3$	$J(\lambda^k)$
1	0.5	0.4	0.1	119.361	114.875	1,093.288	1,093.288
2	0.204426	0.100323	0.695249	251.698	1,283.885	333.333	1,283.885
3	0.058934	0.357712	0.582453	562.026	388.354	558.513	562.026
4	0.094863	0.289738	0.615398	374.660	419.256	511.321	511.321
5	0.049725	0.314217	0.636057	667.014	476.239	500.736	667.014
6	0.085411	0.292701	0.621886	409.412	516.387	507.633	516.387
7	0.063335	0.305152	0.631512	534.273	494.481	503.540	534.273
8	0.069621	0.300798	0.629580	490.292	502.854	504.118	504.118
9	0.067440	0.601731	0.630827	504.600	501.292	503.313	504.600
10	0.067777	0.301340	0.630881	502.326	502.108	503.166	503.166
⋮	⋮	⋮	⋮	⋮	⋮	⋮	⋮
30	0.067709	0.301065	0.631225	502.791	502.799	502.802	502.802
31	0.067708	0.301065	0.631225	502.793	502.799	502.802	502.802
32	0.067708	0.301065	0.631225	502.794	502.764	502.802	502.802
33	0.067708	0.301064	0.631226	502.793	502.802	502.800	502.802
34	0.067708	0.301064	0.631226	502.795	502.802	502.800	502.802
35	0.067708	0.301064	0.631226	502.796	502.801	502.800	502.801

in Figs. 7.1–7.3. They represent a comparison between the trajectories of the original state vector and the trajectories of the estimated states. The estimation error ($e_3^\alpha = x_3^\alpha - \hat{x}_3^\alpha$, $\alpha = 1, 2, 3$) for two different sampling times is graphed in Figs. 7.4 and 7.5. Since the first two components of the state vector are available, then only the third component of the error vector is presented. Figure 7.6 shows a comparison between the control law u_0 when the state vector is completely available and when only the output information is available. This was done for three different sampling times and we can see how by reducing the sampling time the error decreases between the control designed for the nominal system (the state vector is known and there is no uncertainties) and the control using OISM and the hierarchical sliding mode observer. Clearly, this is a consequence of the fact that by reducing the sampling time we reduce the state estimation error.

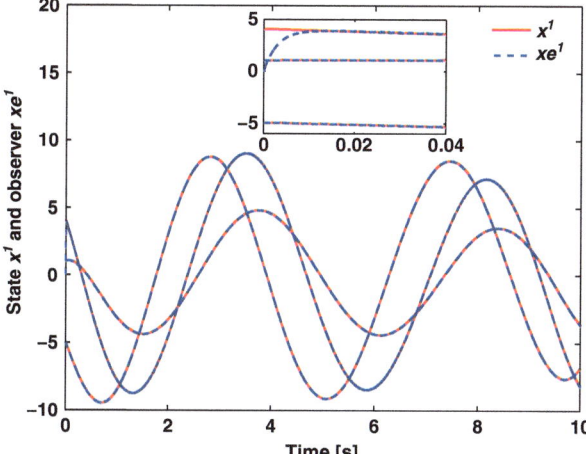

Fig. 7.1. Trajectories of the original state and the estimated one for the first plant.

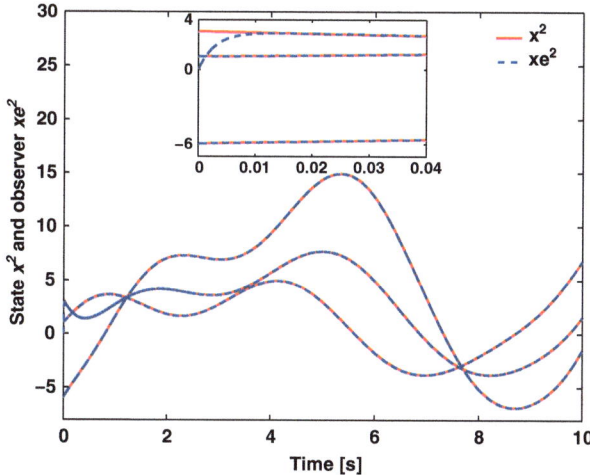

Fig. 7.2. Trajectories of the original state and the estimated one for the second plant.

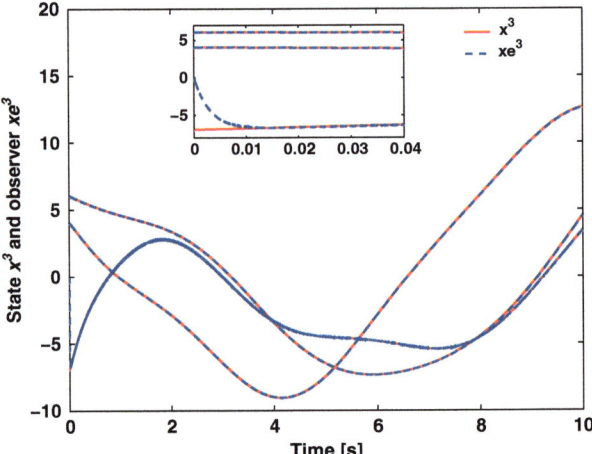

Fig. 7.3. Trajectories of the original state and the estimated one for the third plant.

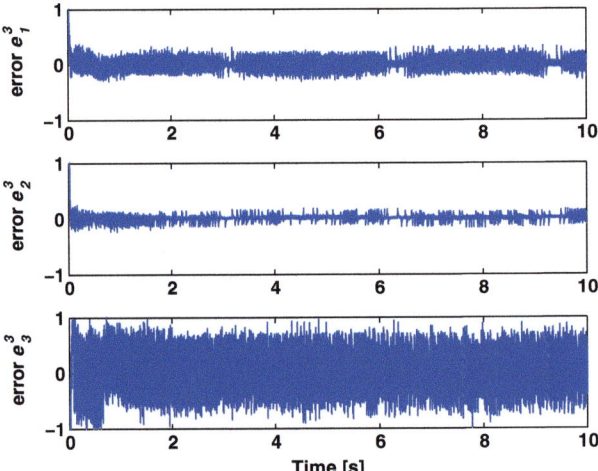

Fig. 7.4. Observation error $e_3^\alpha = x_3^\alpha - \hat{x}_3^\alpha$ for $\delta = 10^{-3}$.

7.7 Example   93

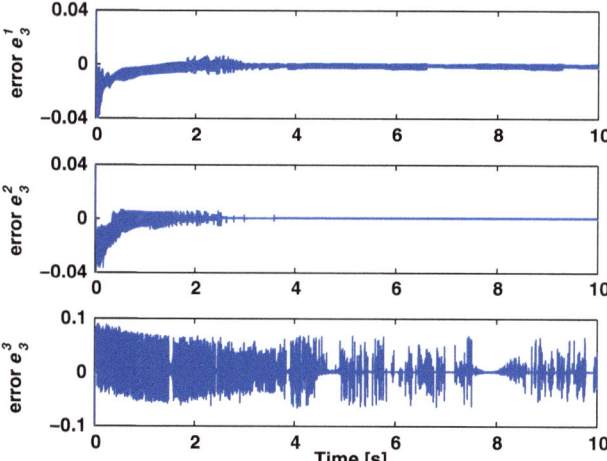

Fig. 7.5. Observation error $e_3^\alpha = x_3^\alpha - \hat{x}_3^\alpha$ for $\delta = 10^{-4}$.

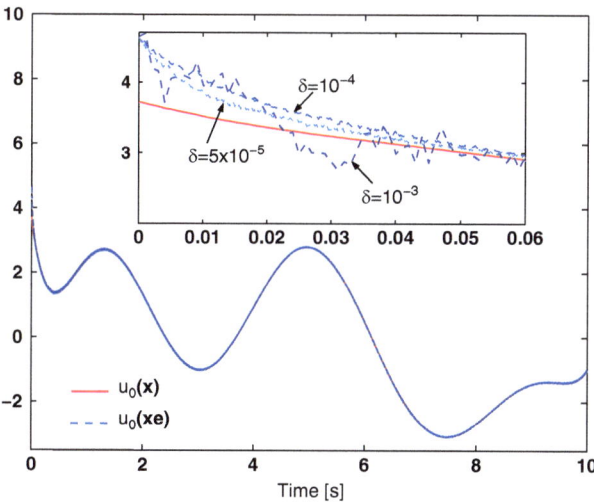

Fig. 7.6. For three different sampling times (δ), a comparison between $u_0(\mathbf{x})$ and $u_0(\hat{\mathbf{x}})$.

Part III

PRACTICAL EXAMPLES

8
Fault Detection

Abstract Here, we will use the OISM to design a fault estimator. We will continue using the cart–pendulum system as in Chap. 4. Here the matched uncertainties are assumed to represent a signal indicating the level of the control performance, that is, in the absence of actuator failures, the matched disturbances do not exist, and when the actuator has a fail, the disturbances appeared in the system. Thus, after the observer is designed and the control law is given, a third step is followed to estimate the fault signal by using the equivalent control method and a low-pass filter.

8.1 Model Description

Let us take again the linearized model of an inverted pendulum over an inverted cart–pendulum given in Fig. 8.1. The aim here is to do a fault estimation allowing to indicate the level of the actuator failure, if it exists. The equations governing the dynamics of the system are as follows:

$$\begin{aligned}\dot{x}(t) &= Ax(t) + Bu + B\gamma(t) \\ y(t) &= Cx(t)\end{aligned} \qquad (8.1)$$

The state vector x consists of four state variables: x_1 is the distance between a reference point and the center of inertia of the trolley; x_2 represents the angle between the vertical and the pendulum; x_3 represents the linear velocity of the trolley; finally, we have that x_4 is equal to the angular velocity of the pendulum. Matrices A, B, and C are

$$A = \begin{bmatrix} 0 & 0 & 1 & 0 \\ 0 & 0 & 0 & 1 \\ 0 & 1.2586 & 0 & 0 \\ 0 & 7.5514 & 0 & 0 \end{bmatrix}, \; B = \begin{bmatrix} 0 \\ 0 \\ 0.1905 \\ 0.1429 \end{bmatrix}, \; C = \begin{bmatrix} 1 & 0 & 0 & 0 \\ 0 & 1 & 0 & 0 \end{bmatrix}$$

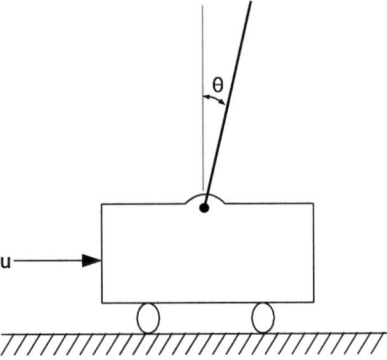

Fig. 8.1. Inverted cart–pendulum.

For this example, we consider that the system output is given by the cart and pendulum positions. We consider that $\gamma(t)$ is given by

$$\gamma(t) = -\alpha(t) u(t), \, \alpha(t) \in [0,1]$$

Hence, $\alpha(t)$ gives an indication of how the actuator is functioning. Thus, if $\alpha(t)$ is equal to zero, the control is in perfect conditions; in the other extreme, if $\alpha(t)$ is equal to one, then the control is totally failing. Then system (8.1) may be rewritten in the following form:

$$\begin{aligned} \dot{x}(t) &= Ax(t) + Bu - B\alpha(t)u(t) \\ y(t) &= Cx(t) \end{aligned} \qquad (8.2)$$

The control law is designed as a infinite horizon LQ optimal control:

$$u^*(t) = \min_{u \in U_{\text{adm}}} \int_0^\infty x^\top(t) Q x(t) + u^\top(t) R u(t) \, dt$$

For this example $Q = 10I$, $R = 1$. Thus, $u^*(t) = -Kx(t)$, where

$$K = \begin{bmatrix} -3.16 & 158.56 & -9.26 & 59.12 \end{bmatrix}$$

Since $x(t)$ is not directly available, we will use an estimate of it $\hat{x}(t)$, i.e., $u(t) = -K\hat{x}(t)$, where $\hat{x}(t)$ will be designed following the procedure given in Chap. 3.

8.2 Observer Design

Let $\tilde{x}(t)$ be defined by the following dynamic equation,

$$\dot{\tilde{x}}(t) = A\tilde{x}(t) + Bu(t) + L(y(t) - C\tilde{x}(t)) \qquad (8.3)$$

8.2 Observer Design

where L is designed so that $\hat{A} := (A - LC)$ be Hurwitz. For this example L was chosen as

$$L = \begin{bmatrix} 0.6 & 0 \\ 0 & 0.4 \\ 0.08 & 1.25 \\ 0 & 7.58 \end{bmatrix}$$

We will assume that the initial condition for $x(t)$ is a bounded region, that is, $\|x(0)\| \leq 10$. Thus, we recall that the norm of $r(t) = x(t) - \tilde{x}(t)$ stays bounded [see (3.6)].

In this case it is enough to reconstruct the vector $CAx(t)$. Hence, to recover $CAx(t)$, let us introduce an *auxiliary state vector* $x_{\text{a}}^{(1)}(t)$ governed by

$$\dot{x}_{\text{a}}^{(1)}(t) = A\tilde{x}(t) + Bu + \tilde{L}\left(C\tilde{L}\right)^{-1} v^{(1)}(t) \tag{8.4}$$

where $\tilde{L}$ satisfies $\det C\tilde{L} \neq 0$; in this case $\tilde{L}$ is designed as $C^\top$ and so $CC^\top = I$. Furthermore, we chose $x_{\text{a}}^{(1)}(0)$ so that

$$Cx_{\text{a}}^{(1)}(0) = y(0)$$

The variable $s^{(1)} \in \mathbb{R}^p$ is defined as

$$s^{(1)}\left(y(t), x_{\text{a}}^{(1)}(t)\right) = y(t) - Cx_{\text{a}}^{(1)}(t) \tag{8.5}$$

Then, since $CB = 0$, the derivative of s along the time is

$$\dot{s}^{(1)}(t) = CA\left(x(t) - \tilde{x}(t)\right) - v^{(1)}(t) \tag{8.6}$$

with the output injection $v(t)$ given by:

$$v^{(1)} = \begin{cases} M(t) \dfrac{s^{(1)}}{\|s^{(1)}\|} & \text{if } s^{(1)} \neq 0 \\ 0 & \text{if } s^{(1)} = 0 \end{cases}$$

Here the gain scalar function $M_1(t)$ should satisfy the condition

$$M_1(t) > \|CA\| \, \|r(t)\| \tag{8.7}$$

to obtain the sliding mode regime. Thus from (8.6), the equivalent output injection is

$$v_{\text{eq}}^{(1)}(t) = CAx(t) - CA\tilde{x}(t)$$
$$= \begin{bmatrix} x_3(t) \\ x_4(t) \end{bmatrix} - \begin{bmatrix} \tilde{x}_3(t) \\ \tilde{x}_4(t) \end{bmatrix}, \forall t > 0$$

Thus, $CAx(t)$ is reconstructed by means of the following representation:

$$\begin{bmatrix} x_3(t) \\ x_4(t) \end{bmatrix} = \begin{bmatrix} \tilde{x}_3(t) \\ \tilde{x}_4(t) \end{bmatrix} + v_{\text{eq}}^{(1)}(t), \forall t > 0 \tag{8.8}$$

8.3 Fault Estimation

Let us designed a second auxiliary system with $x_a^{(2)}(t)$ generated by

$$\dot{x}_a^{(2)}(t) = A^2\tilde{x}(t) + ABu(t) + C^\top v^{(2)}(t)$$

where $x_a^{(2)}(0)$ satisfies

$$v_{eq}^{(1)}(0) + CA\tilde{x}(0) - Cx_a^{(2)}(0) = 0$$

As for $s^{(2)} \in \mathbb{R}^p$, it is defined by

$$s^{(2)}\left(v_{eq}^{(1)}(t), x_a^{(2)}(t)\right) = CA\tilde{x}(t) + v_{eq}^{(1)}(t) - Cx_a^{(2)}(t)$$

The output injection $v^{(2)}(t)$ is

$$v^{(2)} = \begin{cases} M_2(t) \dfrac{s^{(2)}}{\|s^{(2)}\|} & \text{if } s^{(2)} \neq 0 \\ 0 & \text{if } s^{(2)} = 0 \end{cases} \quad (8.9)$$
$$M_2(t) > \|CA^2\| \|r(t)\|$$

Thus $CA^2x(t)$ can be recovered by means of the equality:

$$CA^2x(t) - CAB\alpha(t)u(t) = CA^2\tilde{x}(t) + v_{eq}^{(2)}(t), \quad t > 0 \quad (8.10)$$

Now, from (8.10), we have that

$$CAB\alpha(t)u(t) = CA^2x(t) - CA^2\tilde{x}(t) - v_{eq}^{(2)}(t)$$

In our example

$$CAB = \begin{bmatrix} 0.1905 \\ 0.1429 \end{bmatrix}, \quad CA^2x(t) = \begin{bmatrix} 0 & 1.2586 \\ 0 & 7.5514 \end{bmatrix} \begin{bmatrix} x_1 \\ x_2 \end{bmatrix}$$

For the case $u(t) \neq 0$, $\alpha(t)$ can be expressed as

$$\alpha(t)u = \begin{bmatrix} 0.1905 \\ 0.1429 \end{bmatrix}^+ \left(\begin{bmatrix} 0 & 1.2586 \\ 0 & 7.5514 \end{bmatrix}\left(y(t) - \begin{bmatrix} \tilde{x}_1 \\ \tilde{x}_2 \end{bmatrix}\right) - v_{eq}^{(2)}(t)\right)$$

As explained in Chap. 3, $v_{eq}^{(1)}(t)$ and $v_{eq}^{(2)}(t)$ are recovered by using low-pass filters, i.e.,

$$\tau \dot{v}_{av}^{(k)}(t) + v_{av}^{(k)}(t) = v^{(k)}(t); \quad v_{av}^{(k)}(0) = 0, \, k = 1, 2$$

Thus the estimation of $\alpha(t)u$ is carried out by means of

$$\widehat{\alpha u} = \begin{bmatrix} 0.1905 \\ 0.1429 \end{bmatrix}^+ \left(\begin{bmatrix} 1.2586 \\ 7.5514 \end{bmatrix}(y_2(t) - \tilde{x}_2) - v_{av}^{(2)}(t)\right)$$

For the simulations we use a sample step of 2×10^{-5}. In Fig. 8.2, the function $\alpha(t)$ used for the simulations is shown. The comparison of the values of $\alpha(t)u(t)$ and its estimation $\widehat{\alpha u}$ is given in Fig. 8.3

8.3 Fault Estimation   101

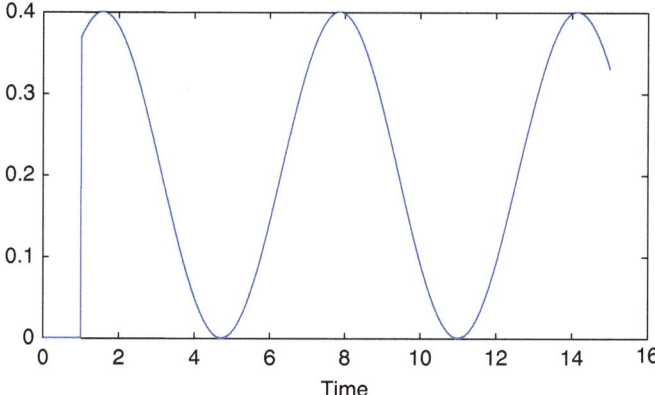

Fig. 8.2. Index of the failure ($\alpha(t)$).

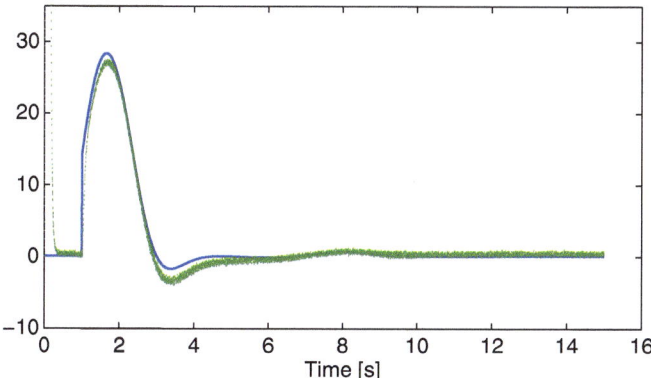

Fig. 8.3. Comparison of $\alpha(t)u(t)$ (*solid*) and its estimation $\widehat{\alpha u}$ (*dotted*).

9
Stewart Platform

Abstract We present an application associated with the so-called Stewart's platform, which is a robot of closed cinematic chain. This is one of the most important examples of totally parallel manipulator, understanding as such a robot that possess two bodies, one fixed and the other mobile, which are connected between them by several arms. Typically each arm is controlled by an actuator. Stewart's platform has, therefore, a parallel configuration of six degrees of freedom composed of two rigid bodies connected by six prismatic actuators. The biggest rigid body is named the base, and the mobile body is called the mobile platform. Here, the goal is to design a robust control to stabilize Stewart's platform with three degrees of freedom around a wished position when we do not have complete information with regard to the initial conditions and the permanent disturbance that affect this platform. The specific application consists in an aerostatic balloon easy to manipulate that is mooring to earth by a cable of approximately 400 m of length. This balloon is connected to the base platform and a video camera is fixed to the mobile platform to keep under surveillance a specific area of $20\,\mathrm{km}^2$ approximately. Since the platform basis is over the mobile one, we will name this platform *inverted Stewart's platform*. Due to the type of application, the platform is permanently under the action of the force of the wind. Therefore, we will work with the wind's acceleration as our permanent disturbance. Another characteristic of our implementation is that we have only output (not state) information available. In this situation the implementation of an OISM control seems to be useful.

9.1 Model Description

Let us describe the inverted Stewart's platform, which consists of a base platform and a mobile one, both with shape of an equilateral triangle, of sides a and b, $a > b$, respectively. The vertexes of the base are joined to

104   9 Stewart Platform

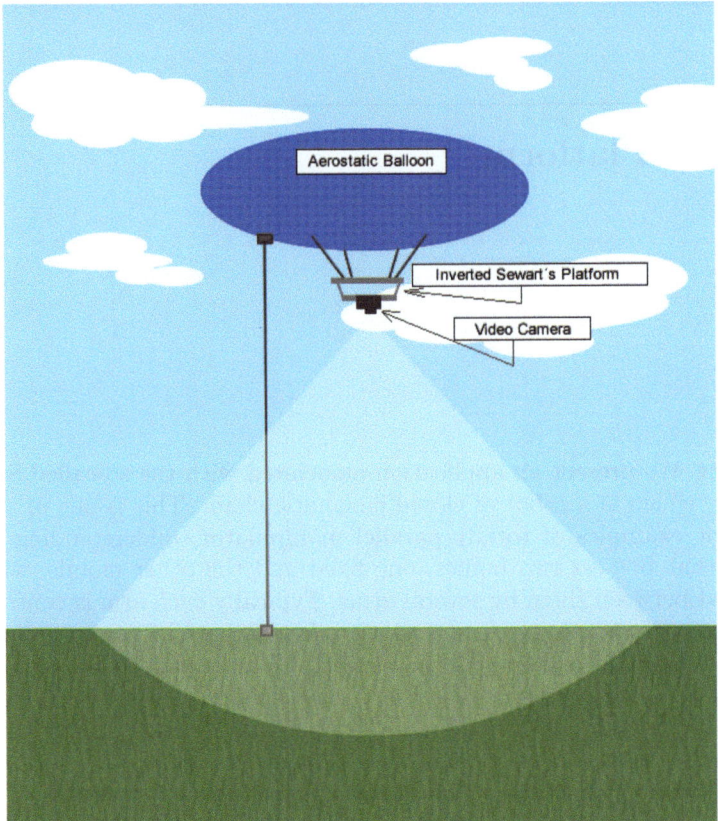

Fig. 9.1. Scheme of the remote surveillance device.

the correspondent vertexes of the mobile platform by actuators of lengths, l_i ($i = 1, 2, 3$), variable and enclosed. These actuators are fastened to the base platform in the points A_i ($i = 1, 2, 3$) by cylindrical joints which axes of rotation perpendiculars to the segment $\overline{A_i A_0}$ ($i = 1, 2, 3$). And they are connected to the mobile platform in the respective points B_i ($i = 1, 2, 3$) by spherical joints (see Fig. 9.2). The type of joints used to connect the platforms, base and mobile, through the actuators allow to restrict the six original degrees of freedom to three: two rotations (α and β) and one translation (h). Two sensors measure the angular velocities, concerning to both rotations α and β of the mobile platform regarding to the horizontal plane and a GPS to measure the position that will allow us to recover the mentioned angles. We are going to assume that the measurement error of the GPS is small enough not to be taken in account in our application. The wished position to stabilize is $\alpha = \beta = 0$ and $h = h_0$.

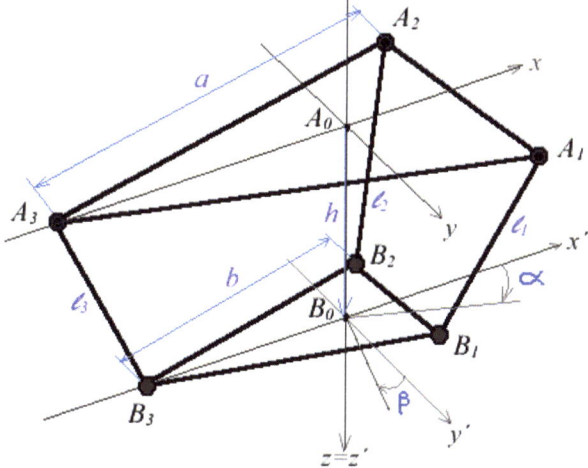

Fig. 9.2. Geometric scheme of the platform P.

This way we obtain the following linear time invariant model with uncertainties of the platform P [58],

$$\dot{x}(t) = Ax(t) + B(u(t) + \gamma(x,t)); \quad x(0) = x^0 \\ y(t) = Cx(t) \tag{9.1}$$

where $x(t) \in \mathbb{R}^6$ is the state vector, $u(t) \in \mathbb{R}^3$ is the control law, $y(t) \in \mathbb{R}^5$ is the output of the system, and w is the permanent perturbation, which represents the wind's acceleration. The vector state x consists of six state variables: $x_1 = \alpha - \alpha_0$, $x_3 = \beta - \beta_0$, $x_5 = (h - h_0)/h_0$, and x_2, x_4, and x_6 represent the velocity of x_1, x_3, and x_5, respectively. There exist two kinds of influences of the external disturbance on the platform P, which are known as general (normal) resonance and parametric resonance. The most important analysis is when the parametric resonance occurs, due to the fact that this one is more dangerous because it grows exponentially, whereas the normal resonance grows linearly. That's why, according to the supposition that the coefficient of the additional presence is small, we are going to study the parametric influence. In Sect. 9.4 we will include an additional influence in the simulation and verify how the system is affected.

The matrices A, B, and C are described below.

$$A = \begin{pmatrix} 0 & 1 & 0 & 0 & 0 & 0 \\ \frac{b^2 \cos^2 \gamma_0 - b(a-b)}{6r_y^2} & 0 & 0 & 0 & 0 & 0 \\ 0 & 0 & 0 & 1 & 0 & 0 \\ 0 & 0 & -\frac{b^2 \cos^2 \gamma_0 + b(a-b)}{6(h_0^2 + r_x^2)} & 0 & 0 & 0 \\ 0 & 0 & 0 & 0 & 0 & 1 \\ 0 & 0 & 0 & 0 & -\cos^2 \gamma_0 & 0 \end{pmatrix} \tag{9.2}$$

$$B = \begin{pmatrix} 0 & 0 & 0 \\ -\frac{bh_0}{6\sqrt{3}r_y^2} & -\frac{bh_0}{6\sqrt{3}r_y^2} & \frac{bh_0}{3\sqrt{3}r_y^2} \\ 0 & 0 & 0 \\ \frac{bh_0}{6(h_0^2+r_x^2)} & -\frac{bh_0}{6(h_0^2+r_x^2)} & 0 \\ 0 & 0 & 0 \\ -\frac{1}{3} & -\frac{1}{3} & -\frac{1}{3} \end{pmatrix} \quad (9.3)$$

and

$$C = \begin{pmatrix} 1 & 0 & 0 & 0 & 0 & 0 \\ 0 & 1 & 0 & 0 & 0 & 0 \\ 0 & 0 & 1 & 0 & 0 & 0 \\ 0 & 0 & 0 & 1 & 0 & 0 \\ 0 & 0 & 0 & 0 & 1 & 0 \end{pmatrix} \quad (9.4)$$

As in Chap. 4, now, for the system (9.1), we design the control law u to be

$$u = u_0 + u_1 \quad (9.5)$$

where the control $u_0 \in \mathbb{R}^m$ is the ideal control designed for the nominal system (i.e., $\gamma = 0$) and $u_1 \in \mathbb{R}^m$ is designed to compensate the uncertainty $\gamma(x,t)$ from the initial time.

9.2 Output Integral Sliding Mode

According to (4.8) and (4.6) control u_1 is designed in the following form:

$$u_1 = -\beta(t) \frac{s(t)}{\|s(t)\|} \quad (9.6)$$
$$\beta(t) - \left(\gamma^+(y,t) + \left\|(CB)^+ CA\right\| \|x(t) - \hat{x}(t)\|\right) \geq \lambda > 0$$

with

$$s(y(t)) = (CB)^+ y(t) - \int_0^t \left((CB)^+ CA\hat{x}(\tau) - u_0(\tau)\right) d\tau - (CB)^+ y(0) \quad (9.7)$$

The observer in this case is designed as follows:

$$\hat{x}(t) = \tilde{x}(t) + H^+ v_{\text{av}}(t)$$
$$v_{\text{av}} = \left[\left(Cx_a^{(1)} - C\tilde{x}(t)\right)^T \left(v_{\text{av}}^{(1)}\right)^T\right]^T \quad (9.8)$$

with

$$\dot{\tilde{x}}(t) = \tilde{A}\tilde{x}(t) + Bu_0(t) + B(CB)^+ CA\hat{x}(t) + L(y(t) - C\tilde{x}(t)) \quad (9.9)$$

where L must be designed such that the eigenvalues of $\hat{A} := (\tilde{A} - LC)$ have negative real part. $H = \begin{bmatrix} C \\ C\tilde{A} \end{bmatrix}$ and v_{av} is calculated as follows

$$\tau \dot{v}_{\text{av}}^{(1)}(t) + v_{\text{av}}^{(1)}(t) = v^{(1)}(t); \quad v_{\text{av}}^{(1)}(0) = 0$$

with $v^{(1)}$ designed as

$$v^{(1)} = \begin{cases} M_1 \dfrac{s^{(1)}}{\|s^{(1)}\|} & \text{if } s^{(1)} \neq 0 \\ 0 & \text{if } s^{(1)} = 0 \end{cases}$$

with $s^{(1)} \in \mathbb{R}^5$ defined by

$$s^{(1)}\left(y(t), x_{\text{a}}^{(1)}(t)\right) = Cx(t) - Cx_{\text{a}}^{(1)}(t) \tag{9.10}$$

and $x_{\text{a}}^{(1)}(t)$ takes the form

$$\dot{x}_{\text{a}}^{(1)}(t) = \tilde{A}\tilde{x}(t) + B\left[u_0(t) + (CB)^+ CA\hat{x}(t)\right] + \bar{L}\left(C\bar{L}\right)^{-1} v^{(1)}(t) \tag{9.11}$$

where $\bar{L} \in \mathbb{R}^{6 \times 5}$ is a matrix so that $\det(C\bar{L}) \neq 0$ and $x_{\text{a}}^{(1)}(0)$ satisfies

$$Cx_{\text{a}}^{(1)}(0) = y(0)$$

9.3 Min–Max Stabilization of Platform P

Let us consider for our application the nominal control u_0 as a control with linear output feedback:

$$u_0 = Ky \tag{9.12}$$

where $K \in \mathcal{F} = \{F \subset \mathbb{R}^{m \times p} | Re(\lambda_i) \leq -k_0, k_0 > 0\}$ and λ_i ($i = 1, \cdots, n$) are the eigenvalues of the matrix $A_1(K) := A + BKC$.

In the ideal sliding motion, the dynamic equations for the state x have the form

$$\dot{x}(t) = A_1(K)x(t), \ x(0) = x^0 \tag{9.13}$$

Thus, the *min–max problem* consists in finding the values of k_{ij} which satisfy the following evaluation criterium

$$J(K) = \max_{|x(0)| \leq \mu} \int_0^\infty x^T(t)Qx(t)\,dt \to \min_{K \in \mathcal{F}} \tag{9.14}$$

where $Q = Q^\top \geq 0$; we chose Q as the identity matrix of dimension n.

Physically this means that given the worst initial conditions the control minimizes the deviations in time of the system parameters and also in this way an asymptotically stable behavior is achieved. For our application, as a remote surveillance device, it is of great importance to decrease not only the angle deviations but also their velocities since we need the movement of the camera to be slow enough to capture better images.

Thus, the control law solving (9.14) for (9.13) is of the form

$$u_0(t) \equiv u_0^*(t) = K^* y(t)$$

Let us change the optimal control problem (9.14) to a nonlinear programming problem (see [59]). For that let us consider the differential matrix equation:

$$\dot{Z} = A_1^T Z + Z A_1, \quad Z(0) = Q \tag{9.15}$$

The general solution of (9.15) has the form

$$Z(t) = e^{A_1^T t} Q e^{A_1 t} \tag{9.16}$$

Since, for any $K \in \mathcal{F}$, A_1 matrix is stable, then the integral $\int_0^\infty Z(t) dt$ converges. Thus we have

$$A_1^T \int_0^\infty Z(t) dt + \int_0^\infty Z(t) dt A_1 = \int_0^\infty \dot{Z}(t) dt =$$
$$= Z(\infty) - Z(0) = -F$$

Then it is possible to affirm that the matrix

$$P = \int_0^\infty Z(t) dt \tag{9.17}$$

is the solution of the matrix equation

$$A_1^T P + P A_1 = -F \tag{9.18}$$

As we mentioned before $F = I_n$ (I_n denotes the identity matrix with dimension n). (9.18) is the well-known Lyapunov equation, and its solution is a symmetrical positive-defined matrix. Since A_1 depends on the choice of K matrix, P matrix also depends on K.

Then, the functional $J(\tilde{K})$ can be rewritten as:

$$\max_{|x(0)| \leq \mu} \int_0^\infty x^T x \, dt = \max_{|x(0)| \leq \mu} \int_0^\infty x^T(0) Z(t) x(0) \, dt = \max_{|x(0)| \leq \mu} x^T(0) P(K) x(0) \tag{9.19}$$

On the other hand, for any symmetrical definite positive matrix, it fulfills the following inequality:

$$x^T(0) P(K) x(0) \leq \lambda_{max}(P(K)) \mu^2 \tag{9.20}$$

Between all the initial conditions, $|x(0)| \leq \mu$, there exists one for which the equality is reached in (9.20). Consequently the functional can be expressed in the following way:

$$\max_{|x(0)| \leq \mu} \int_0^\infty x^T(t) x(t) dt = \mu^2 \lambda_{\max}(P(K))$$

This way we can reduce the min–max problem (9.14) to the following extrema problem of finite dimension:

$$\mu^2 \lambda_{\max}(P(K)) \to \min_{K \in \mathcal{F}} \quad (9.21)$$

Let K^* be the matrix solving the optimization problem (9.21).

$$u_0(t) = K^* y(t) \quad (9.22)$$

9.4 Numerical Simulations

Let us consider the following structural dimensions for our platform P: $a = 0.5\,\mathrm{m}$; $b = 0.3\,\mathrm{m}$; $g = 9.81\,\mathrm{m/s^2}$; $h_0 = 0.2\,\mathrm{m}$; $\gamma_0 = 60°$, and $m = 3\,\mathrm{kg}$ (see Fig. 9.2). Then, the matrices A and B for the motion equations (9.1) are

$$A = \begin{pmatrix} 0 & 1 & 0 & 0 & 0 & 0 \\ -1.875 & 0 & 0 & 0 & 0 & 0 \\ 0 & 0 & 0 & 1 & 0 & 0 \\ 0 & 0 & -0.3433 & 0 & 0 & 0 \\ 0 & 0 & 0 & 0 & 0 & 1 \\ 0 & 0 & 0 & 0 & -0.25 & 0 \end{pmatrix}$$

and

$$B = \begin{pmatrix} 0 & 0 & 0 \\ -1.732 & -1.732 & 3.464 \\ 0 & 0 & 0 \\ 0.2105 & -0.2105 & 0 \\ 0 & 0 & 0 \\ -\frac{1}{3} & -\frac{1}{3} & -\frac{1}{3} \end{pmatrix}$$

The vector $\gamma(w, x, t)$ is

$$\gamma(w, x, t) = \begin{pmatrix} wx_3 + 1.3686 wx_5 \\ wx_3 - 1.3686 wx_5 \\ wx_3 \end{pmatrix} \quad (9.23)$$

We assume that $w(t)$ represents the wind acceleration, and it takes the following expression $w(t) = 0.1 + 0.5 \sin t$.

Notice that in (9.23) the perturbation only affects the deviation of parameters β and h. This occurs because we assume that the wind only acts on the direction of axis y (see Fig. 9.2). The wished point that we want to stabilize is $(0, 0, 0, 0, h_0, 0)$.

Nevertheless, for our application, the changes in the height of the center of mass of the mobile platform with regard to the plane of the platform base are not of vital importance, since the platform P is set at a height of approximately 400 m of the level of the ground and we wish to maintain the horizontal position of the mobile platform in order to keep certain area under surveillance. That's why we are going to focus on the behavior of the states x_1, x_2, x_3, and x_4, corresponding to the deviation of α and β and their velocities.

The matrix $\tilde{A} = [I - B(CB)^+C]A$ takes the form

$$\tilde{A} = \begin{pmatrix} 0 & 1 & 0 & 0 & 0 & 0 \\ -0.2512 & 0 & 0 & 0 & 0 & 0 \\ 0 & 0 & 0 & 1 & 0 & 0 \\ 0 & 0 & -0.0023 & 0 & 0 & 0 \\ 0 & 0 & 0 & 0 & 0 & 1 \\ -0.1563 & 0 & 0 & 0 & -0.25 & 0 \end{pmatrix}$$

Control u_0 is taken as

$$u_0 = \begin{pmatrix} 3.53x_1 + 5x_2 \\ 4.982x_3 + 9.982x_4 \\ 7.059x_1 + 9.999x_2 + 5.2369x_3 + 7.4865x_4 + 3.36x_5 \end{pmatrix}$$

Matrix L is selected as $C^\top$.

Different simulations were carried out, each one with a sampling step of $\Delta = 2 \cdot 10^{-3}$ and $\Delta = 2 \cdot 10^{-4}$, respectively. The filter constant, τ, was chosen as $\tau = \Delta^{1/2}$. The trajectories of the state vector corresponding to the behavior of the deviation of x_i ($i = 1, \cdots, 4$) under the perturbation w when we use only nominal control u_0 and when we also use control u_1 are compared in Figs. 9.3–9.4 and 9.5–9.6.

We also see the observation error $e(t) = x(t) - \hat{x}(t)$ in Fig. 9.7. As expected, we can see in those figures that the convergence to zero is better when Δ is smaller.

Now, we include in (9.1) the additional influence of the perturbation w given by the following expression

$$\gamma(w, x, t) = \begin{pmatrix} wx_3 + 1.3686wx_5 + \frac{1}{0.2105}\left(\frac{h_0^2}{h_0^2 + r_x^2}\right)w \\ wx_3 - 1.3686wx_5 - \frac{1}{0.2105}\left(\frac{h_0^2}{h_0^2 + r_x^2}\right)w \\ wx_3 \end{pmatrix}$$

$$\tilde{g} = \begin{bmatrix} 0 & 0 & 0 & -\left(\frac{h_0^2}{h_0^2 + r_x^2}\right)w & 0 & 0 \end{bmatrix}^\top$$

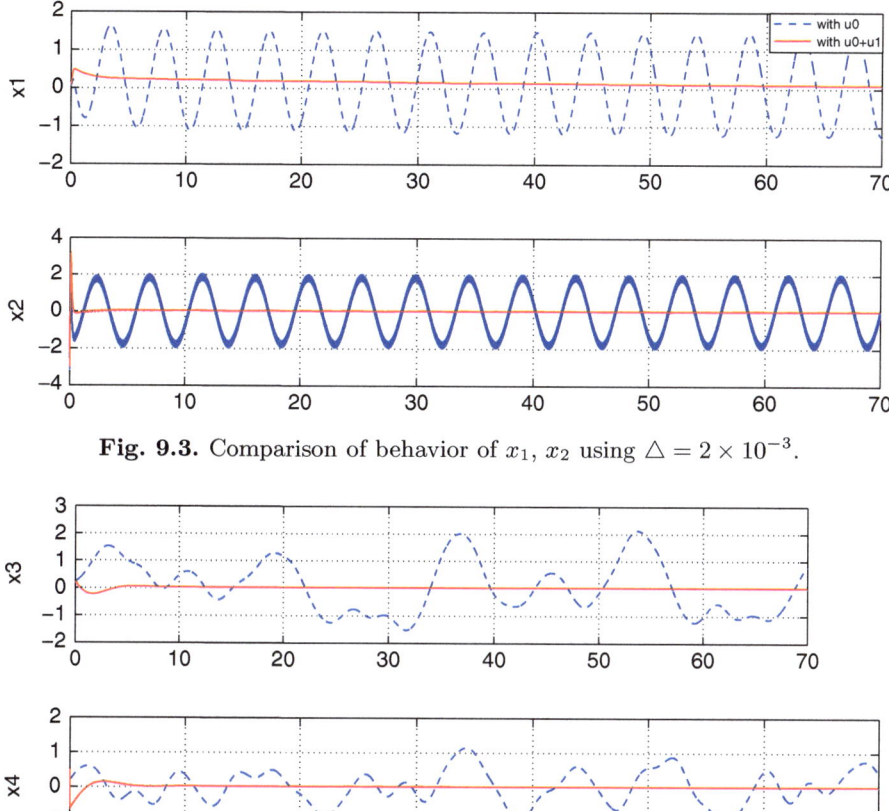

Fig. 9.3. Comparison of behavior of x_1, x_2 using $\Delta = 2 \times 10^{-3}$.

Fig. 9.4. Comparison of behavior of x_3 and x_4 using $\Delta = 2 \times 10^{-3}$.

In Fig. 9.8 we compare the behavior of state x_3 using $u_0 + u_1$ (for $\Delta = 2 \cdot 10^{-4}$) when we have only the parametric influence and when we have both influences of the external perturbation w. We see that in the second case there is a slightly bigger oscillation. This is a deviation of 1.22×10^{-5} degrees from the wished position of angle β and it represents a deviation of the video camera of 5 mm in the ground.

112   9 Stewart Platform

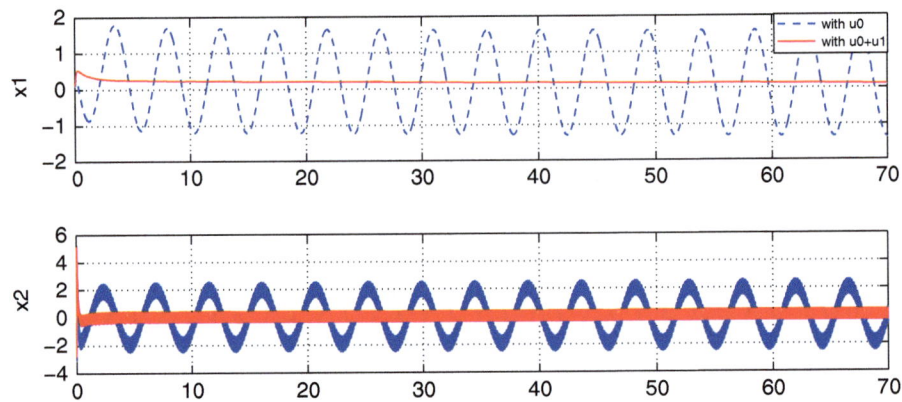

Fig. 9.5. Comparison of behavior of x_1, x_2 using $\triangle = 2 \times 10^{-4}$.

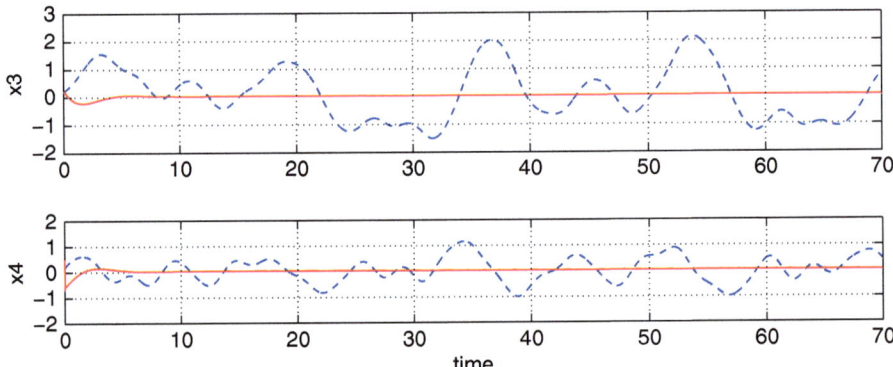

Fig. 9.6. Comparison of behavior of x_3 and x_4 using $\triangle = 2 \times 10^{-4}$.

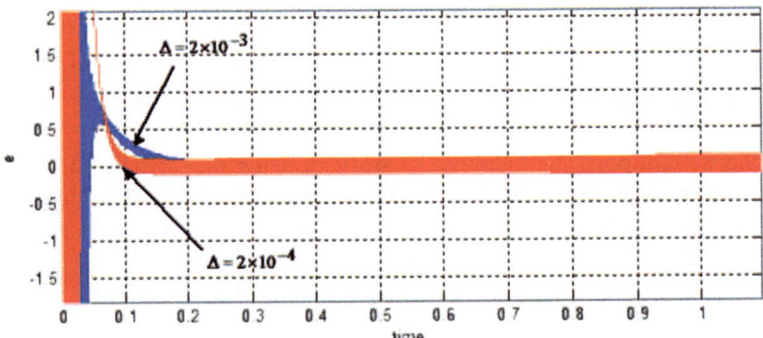

Fig. 9.7. Observation error $e = x_6 - \hat{x}_6$.

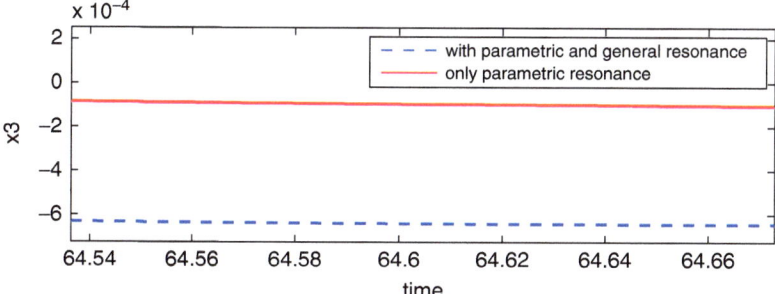

Fig. 9.8. Comparison of behavior of x_3 when the general resonance of w is added.

10
Magnetic Bearing

Abstract Here, we present an example of an application of the output integral sliding mode. We will apply that method to a magnetic levitator. We will consider a magnetic bearing system, which is composed of a planar rotor disk and two sets of stator electromagnets: one acting in the y-direction and the other acting in the x-direction. This system may be decoupled into two subsystems, one for each direction, with similar equations. Here, only the linearized subsystem in the y-direction is considered.

10.1 Preliminaries

The optimal control used for this example is based on an LQ differential game (LQDG) where the players' dynamic is represented by linear ordinary differential equations

$$\dot{x}(t) = Ax(t) + \sum_{i=1}^{2} B^i \left(u^i(t) + \gamma^i(t) \right) \tag{10.1}$$

$$y^1(t) = C^1 x(t), \qquad y^2(t) = C^2 x(t)$$

$$x(0) = x_0, \; t \in [0, t_1]$$

$A \in \mathbb{R}^{n \times n}$ and $B^i \in \mathbb{R}^{n \times m_i}$ ($i = 1, 2$) and $\zeta^i(t) \in \mathbb{R}$ is an unknown input. In addition, $x(t) \in \mathbb{R}^n$ is the game state vector, with $u^i(t) \in \mathbb{R}^{m_i}$ being the control strategies of each i-player, and $y^i(t) \in \mathbb{R}^{p_i}$ is the output of the game for each player which can be measured at each time. Finally, $C^i \in \mathbb{R}^{p_i \times n}$ is the output matrix for player i.

As for the optimal control, let us consider the nominal system, i.e., that γ is identical to zero, then we consider the following quadratic cost functional:

$$J^i(u_0^i, u_0^{\hat{\imath}}) = \int_0^\infty (x^T Q^i(t) x + \sum_{j=1}^{2} u_0^{jT} R^{ij} u_0^j) dt, \; j \neq i \tag{10.2}$$

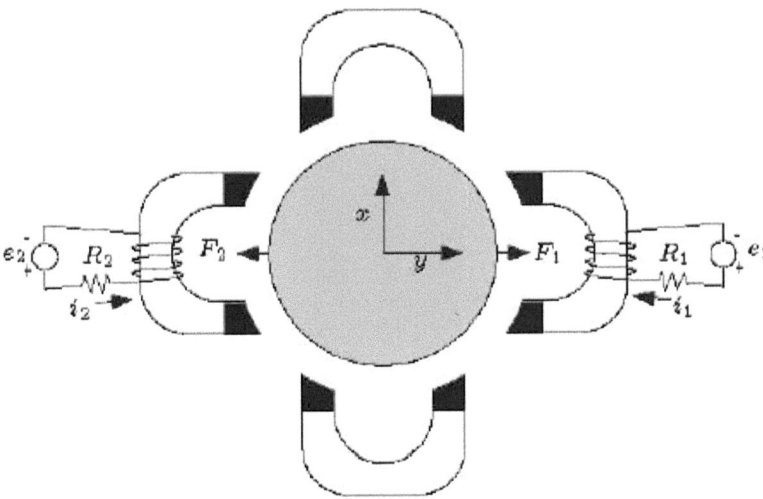

Fig. 10.1. Top view of a planar rotor disk magnetic bearing system [60].

The performance index $J^i(u_0^i, u_0^{\hat{i}})$ (10.2) of each i-player for infinite time horizon nominal game is given in the standard form, where u_0^i is the strategy for i-player and $u_0^{\hat{i}}$ are the strategies for the rest of the players ($\hat{i}$ is the player counteracting to the player with index i). Matrices $Q^i(t)$ and $R^{ji}(t)$ should satisfy the following conditions:

$$Q^i(t) = Q^{i\mathsf{T}}(t) \geq 0, \quad R^{ji}(t) = R^{ji\mathsf{T}}(t) > 0$$
$$R^{ij}(t) = R^{ij\mathsf{T}}(t) \geq 0 \quad (j \neq i) \tag{10.3}$$

Thus, from the limiting solution of the finite-time problem [62], the next coupled algebraic equations appear [63]:

$$-\left(A - S^2 P^2\right)^{\mathsf{T}} P^1 - P^1 \left(A - S^2 P^2\right) + P^1 S^1 P^1 - Q^1 - P^2 S^{21} P^2 = 0 \tag{10.4}$$

$$-\left(A - S^1 P^1\right)^{\mathsf{T}} P^2 - P^2 \left(A - S^1 P^1\right) + P^2 S^2 P^2 - Q^2 - P^1 S^{12} P^1 = 0 \tag{10.5}$$

with

$$S^i = B^i \left(R^{ji}\right)^{-1} B^{i\mathsf{T}}$$
$$S^{ij} = B^i \left(R^{ji}\right)^{-1} R^{ji} \left(R^{ji}\right)^{-1} B^{i\mathsf{T}} \quad \text{for } j \neq i$$

The following result is well established (see [64]): for a 2-player LQDG described by (10.1) with (10.2), let P^i ($i = 1, 2$) be a symmetric stabilizing solution of (10.4) and (10.5).

Note that it is known that (10.4) and (10.5) in general may not be unique [65]; therefore, we consider only a couple of stabilizing strategies.

Taking

$$F^{i*} := \left(R^{ji}\right)^{-1} B^{i\intercal} P^i$$

for $i = 1, 2$, then $\left(F^{1*}, F^{2*}\right)$ is a feedback Nash equilibrium. The limiting stationary (Nash) strategies are

$$u_0^{i*}(t) = -R^{ji^{-1}} B^{i\intercal} P^i x(t) \qquad (10.6)$$

10.2 Disturbances Compensator

Define for each player the next output based sliding function

$$s^i\left(y^i\right) = G^i y^i - \int_0^t \left(G^i C^i A \hat{x}(\tau) + G^i C^i B^i u_0^i(\tau)\right) d\tau - G^i y^i(0) \qquad (10.7)$$

where vector $\hat{x} \in \mathbb{R}^n$ is the observer state vector which is designed following the procedure given in Chap. 4. We define $G^i = D^i \left(C^i B^i\right)^{\perp}$. Let us remark that, with this assignation of the matrix G^i, the term $\left(C^i B^i\right)^{\perp}$ will cancel all terms related to the opposite player. The matrix $D^i \in \mathbb{R}^{m_i \times (p_i - m_i)}$ is assumed so that the following condition $\det\left(G^i C^i B^i\right) \neq 0$ is satisfied.

The time derivative of s^i takes the form

$$\dot{s}^i\left(y^i\right) = G^i C^i A \left(x - \hat{x}\right) + G^i C^i B^i u_1^i(t) + G^i C^i B^i \gamma^i(t) \qquad (10.8)$$

We propose the control $u_1^i(t)$ as follows:

$$u_1^i(t) = \beta(t) \left(W^i\right)^{-1} \frac{s^i(t)}{\|s^i(t)\|} \qquad (10.9)$$
$$W^i := G^i C^i B^i$$

The function $\beta(t)$ should satisfy the inequality

$$\beta(t) > \left\|G^i C^i A\right\| \|x - \hat{x}\| + \gamma^+ \|W^i\|$$

An estimation for $\|x - \hat{x}\|$ may be done following the procedure given in Chap. 4. Thus, an ideal sliding mode is achieved for all $t \geq 0$. This means that from the beginning of the game, the ISM strategy for each player completely compensates the matched uncertainty. The equivalent control which maintains the trajectories on the sliding surface is

$$u_{1eq}^i(t) = -\left(G^i C^i B^i\right)^{-1} G^i C^i A \left(x - \hat{x}\right) - \gamma^i(t)$$

Substitution of the equivalent control in (10.1) yields the sliding mode dynamic

$$\dot{x}(t) = \bar{A}x(t) + \sum_{i=1}^{2} B^i \left(\left(W^i\right)^{-1} G^i C^i A\hat{x}(t) + u_0^i(t)\right) \quad (10.10)$$

$$y^1(t) = C^1 x(t), \qquad y^2(t) = C^2 x(t)$$

where $\bar{A} := A - \sum_{i=1}^{2} \left(W^i\right)^{-1} G^i C^i A$.

10.3 Observer Design

Now, for the design of the observer, we follow the method explained in Sects. 3.3 and 4.5. For the design of the observer, anyone of the outputs can be used. According to the system we are considering, the vector $\tilde{x}(t)$ should be defined in the following way:

$$\dot{\tilde{x}}(t) = \bar{A}\tilde{x}(t) + \sum_{i=1}^{2} B^i \left(\left(W^i\right)^{-1} G^i C^i A\hat{x} + u_0^i(t)\right) + \sum_{i=1}^{2} L^i \left(y^i - C^i \tilde{x}\right) \quad (10.11)$$

Thus, with $r(t) = x - \tilde{x}$, from (10.10) and (10.11) we have

$$\dot{r}(t) = \left(\bar{A} - L^i C^i\right) r(t) = \hat{A} r(t)$$

The vectors $x_a^{(k)}$ are adapted according to the system under consideration

$$\dot{x}_a^{(k)}(t) = \bar{A}\tilde{x} + \sum_{i=1}^{2} B^i \left(\left(W^i\right)^{-1} G^i C^i A\hat{x} + u_0^i(t)\right) + L^i \left(C^i \bar{L}^i\right)^{-1} v^{(1)}(t)$$

where $\bar{L}^i$ is a matrix of the corresponding dimensions such that $\det\left(C^i \bar{L}^i\right) \neq 0$ and $x_a^{(1)}(0)$ satisfies $C^i x_a^{(1)}(0) = y^i(0)$. Besides these two slight modifications, the observer is designed following the procedure given in Sect. 4.5.

Thus, the control $u^i(\hat{x}, t)$ takes the following form:

$$u^i(\hat{x}, t) = -R^{ji^{-1}} B^{iT} P^i \hat{x} - f(t) \left(W^i\right)^{-1} \frac{s^i(t)}{\|s^i(t)\|}, \quad i = 1, 2 \quad (10.12)$$

10.4 Numerical Simulations

The magnetic bearing system has the following dynamic equations [60]:

$$\dot{x} = \underbrace{\begin{bmatrix} 0 & 1 & 0 & 0 \\ \frac{8L_o I_o^2}{mk^2} & 0 & \frac{2L_o I_o}{mk^2} & -\frac{2L_o I_o}{mk^2} \\ 0 & -\frac{2I_o}{k} & -\frac{kR_1}{L_o} & 0 \\ 0 & \frac{2I_o}{k} & 0 & -\frac{kR_2}{L_o} \end{bmatrix}}_{A} x + \underbrace{\begin{bmatrix} 0 \\ 0 \\ \frac{k}{L_o} \\ 0 \end{bmatrix}}_{B^1} (u^1 + \gamma^1) + \underbrace{\begin{bmatrix} 0 \\ 0 \\ 0 \\ \frac{k}{L_o} \end{bmatrix}}_{B^2} (u^2 + \gamma^2)$$

(10.13)

where $k = 2g_o + a$, g_o is the air gap when the rotor is in the position $y = 0$; a is a positive constant introduced to model the fact that the permeability of electromagnets is finite; $L_o > 0$ is a constant which depends on the system construction; I_o is the premagnetization constant; m is the mass of the rotor; and R_1, R_2 are the resistances in the first set of stator electromagnets. The state variables $x = \begin{bmatrix} y & \dot{y} & i_1 - I_o & i_2 - I_o \end{bmatrix}^T$ and the control inputs $u^1 = e_1 - I_o R_1$ and $u^2 = e_2 - I_o R_2$.

Let us take $m = 2\,\text{kg}$, $L_o = 0.3\,\text{mH}$, $I_0 = 60\,\text{mA}$, $R_{1,\dots,4} = 1\,\Omega$, and $k = 0.002\,\text{m}$. With

$$C^1 = \begin{bmatrix} 1 & 0 & 0 & 0 \\ 0 & 0 & 1 & 0 \\ 0 & 0 & 0 & 1 \end{bmatrix}, \quad C^2 = \begin{bmatrix} 1 & 0 & 0 & 0 \\ 0 & 0 & 0 & 1 \end{bmatrix}$$

and the controller parameters $R^{11} = \text{diag}\left(\begin{bmatrix} 1 & 1 \end{bmatrix}\right)$; $R^{22} = \text{diag}\left(\begin{bmatrix} 1 & 1 \end{bmatrix}\right)$; $Q^1 = Q^2 = 50I$; and $R^{12} = R^{21} = 1$. It can be verified that for this system the triplet (A, B^i, C^i) does not have invariant zeros.

The initial condition is taken as $x(0) = \begin{bmatrix} 0.0005 & 0 & 0.06 & 0.06 \end{bmatrix}^T$. The pair $(\overline{A}, C^1)$ is observable. Matrices $\overline{A}$ and L take the following values:

$$\overline{A} = \begin{bmatrix} 0 & 1 & 0 & 0 \\ 530 & 0 & 0.2 & -0.2 \\ 0 & 0 & 0 & 0 \\ 0 & 0 & 0 & 0 \end{bmatrix}, \quad L = \begin{bmatrix} 25 & 0 & 0 \\ 686 & 0.2 & -0.2 \\ 0 & 10.2 & -0.4 \\ 0 & -0.4 & 10.8 \end{bmatrix}$$

The gain L guarantees that the $\hat{A} = A - LC^1$ matrix is Hurwitz. Applying the Lyapunov iteration algorithm [66] we find $F^1 = \begin{pmatrix} 20949 & 901 & 10 & 3 \end{pmatrix}$ and $F^2 = \begin{pmatrix} -20949 & -901 & -3 & 10 \end{pmatrix}$. The uncertainties are $\zeta^1(t) = 2\sin(4t) + 2\cos(2t) + 1$ and $\zeta^2(t) = 3\cos(5t)$. The output ISM gains are $G^1 = \begin{bmatrix} -1 & 1 & 0 \end{bmatrix}$, $G^2 = \begin{bmatrix} -1 & 0 & 1 \end{bmatrix}$, $M_1^1 = -10$, and $M_1^2 = -10$. The simulation integration time was $10\,\mu\text{s}$, i.e., $\Delta = 10\,\mu\text{s}$, and the filter constant was chosen as $\tau = \Delta^{1/2}$. Figure 10.2 shows the feasibility of the Robust Nash methodology; the effects of the perturbations are clearly compensated. The players' performance indexes are shown in Fig. 10.3 and listed in Table 10.1.

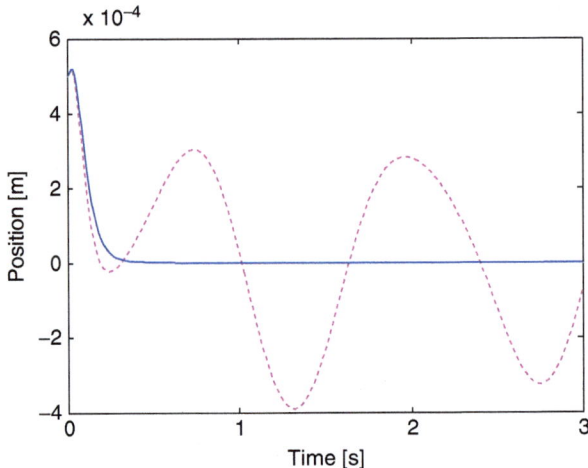

Fig. 10.2. Position of rotor for the perturbed system without compensation (*dotted line*) and using Robust Nash strategy (*solid line*).

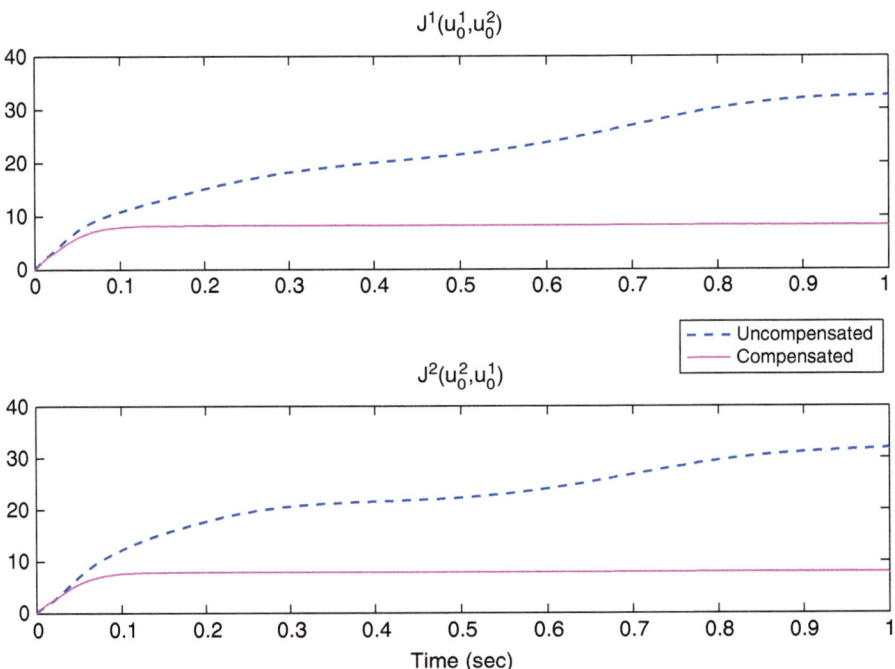

Fig. 10.3. Individual performance index for each player. Perturbed system without compensation (*solid line*) and with compensation (*dash line*).

Table 10.1. Players' performance with and without compensation

t (s)	Nash strategy		Robust Nash strategy	
	$J^1(u_0^1, u_0^2)$	$J^2(u_0^2, u_0^1)$	$J^1(u_0^1, u_0^2)$	$J^2(u_0^2, u_0^1)$
0	0	0	0	0
0.5	22.2	22.7	8.1662	7.7529
1	32.5	31.7	8.1698	7.7565
1.5	48.4	56.4	8.1734	7.7601
2	60.4	65.3	8.177	7.7638
2.5	74.1	81.0	8.1806	7.7673
3	85.8	98.3	8.1842	7.7709
3.5	96.5	107.3	8.1878	7.7745
4	115.4	132.7	8.1914	7.7781

Appendix A
Sliding Modes and Equivalent Control Concept

Abstract This chapter presents basic information about equivalent control method for definition of the conventional sliding mode controllers: Some useful results about online calculation of equivalent control are presented. When using SM control, one of the most interesting and even practical problems appearing is that of finding the trajectory of the state variables, the so-called sliding equations. A formal approach is through the solution of differential inclusions in the Filippov sense. However, a simpler way to study the effect of a discontinuous control acting on a system is the *equivalent control method (ECM)*, which, for affine systems, in fact turns out to give the same results as studying differential inclusions in the Filippov sense. Thus, the aim of this chapter is to introduce a short description of the ECM.

A.1 Introduction

In general, the motion of a control system with discontinuous right-hand side may be described by the differential equation:

$$\dot{x} = f(x,t,u), \quad x \in \mathbb{R}^n, \ u \in \mathbb{R}^m \tag{A.1}$$

$$u_i = \begin{cases} u_i^+(x,t) \text{ if } s_i > 0 \\ u_i^-(x,t) \text{ if } s_i < 0 \end{cases} \text{ for } i = 1, \ldots, m$$

where the vector function $s = s(x)$ defines the sliding manifold $S = \{x : s(x) = 0\}$. It is assumed that $f(x,t,u)$, $u_i^+(x,t)$, $u_i^-(x,t)$, and $s(x)$ are continuous functions of the system state. The motion on the discontinuity surfaces $s_i(x) = 0$ is the so-called *sliding mode motion* (see Fig. A.1). This motion is characterized by high-frequency (theoretically infinite) switching of the control inputs and the fact that, due to changes in the control input, the function $f(x,t,u)$ on the different side of the discontinuity surface $(x^1 \neq x^2)$ satisfies the relation $f(x^1,t,u_i^+) \neq f(x^2,t,u_i^-)$ and consequently conditions for the uniqueness of the solution of the ordinary differential equation do not

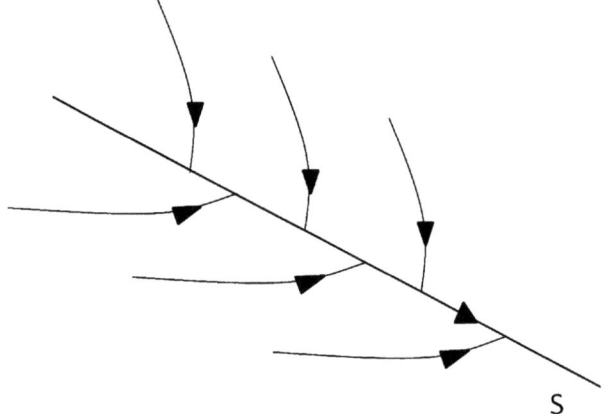

Fig. A.1. Sliding motion.

hold in this case. It has been shown that, if a regularization method yields an unambiguous result, the motion equations on the discontinuity surfaces exist. A regularization method consists in replacing the ideal motion equations (A.1) by more accurate ones $f(x, t, \tilde{u})$. These new equations take into account nonidealities (like hysteresis, delay, etc.) in the implementation of the control input $\tilde{u}$. The new equations have solutions in the conventional sense, but nevertheless motion is no longer restricted to the manifold S but instead evolves in some vicinity Δ (boundary layer) of the manifold. If Δ tends to zero the motion in the boundary layer tends to the motion of system with the ideal control. Equations of motion obtained as results of such a limit process will be regarded as ideal sliding modes. For systems linear with respect to the control input, regularization allows for substantiation of the so-called equivalent control method which is used as a simple procedure for finding the sliding mode motion equations.

A.2 Equivalent Control Method

Consider the system described by the following affine system:

$$\dot{x}(t) = f(x, t) + B(x, t) u(t), \ t \geq t_0 \qquad (A.2)$$

where $x \in \mathbb{R}^n$ and $u \in \mathbb{R}^m$ represent the state vector and the control vector, respectively. Moreover, $f(x, t)$ and $B(x, t)$ are continuous vector and matrix functions, respectively, with respect to all the arguments. Here, u is to be designed as a discontinuous control so as to drive the trajectories of (A.2) into the sliding manifold S and to maintain them there for all future time. The function $s(x) \in \mathbb{R}^m$, which we will call the sliding variable, is to be designed according to some specific requirements. Once the trajectories of (A.2) are in the manifold S, i.e., $s(x) = 0$, we say that (A.2) is on a sliding mode (SM). A u achieving the SM will be referred to as a sliding mode control.

Let us assume that if $s(x(t)) \equiv 0$, then its derivative will also be identical to zero. Thus, we have that

$$\dot{s}(x) = \frac{\partial s}{\partial x}\left[f(x,t) + B(x,t)u(t)\right] = 0 \tag{A.3}$$

Assuming that $G(x) := \frac{\partial s}{\partial x}$ fulfills the condition $\det G(x)B(x) \neq 0$, and $u(t)$ taken from (A.3) is the so-called equivalent control, then we have that

$$u_{eq}(t) = -\left[G(x)B(x,t)\right]^{-1}\left[G(x)f(x,t)\right] \tag{A.4}$$

What the EC method asserts is that the dynamics of (A.2) can be calculated by substituting u_{eq} in the place of u, i.e., on the sliding mode, the system is governed by the following equations:

$$\dot{x} = f(x,t) - B(x,t)\left[G(x)B(x,t)\right]^{-1}\left[G(x)f(x,t)\right] \tag{A.5}$$

Consider then the following simple scalar example:

$$\dot{x}(t) = ax(t) + bu(t) + \gamma(t) \tag{A.6}$$

where a and $b \neq 0$ are real scalars and $\gamma(t)$ is a disturbance. Let's say that we wish to constrain $x(t)$ to the origin in a finite time and in spite of the lack of knowledge of $\gamma(t)$. This can be achieved by selecting $u = -b^{-1}M(t)\operatorname{sign} x$ and $M(t) > |ax| + |\gamma(t)| + \epsilon$, for some arbitrarily small ϵ. By differentiating $V = \frac{1}{2}|x|^2$ we get

$$\dot{V} = |x|(ax + bu + \gamma) \leq -|x|(M(t) - |ax| - |\gamma|)$$
$$\leq -|x|\epsilon = -\sqrt{2}\epsilon\sqrt{V}$$

By using the comparison principle, we obtain that

$$\frac{|x(t)|}{\sqrt{2}} = \sqrt{V(t)} \leq \sqrt{V(t_0)} - \frac{\epsilon}{\sqrt{2}}(t - t_0) \text{ for all } t \geq t_0 \tag{A.7}$$

Since $V(t)$ is by definition a positive function, from (A.7) we can calculate an upper estimate of the time t_s when $V(t)$ vanishes and consequently $x(t)$ does as well. Consequently, we obtain that

$$t_s \leq \frac{\sqrt{2}}{\epsilon}V(t_0) + t_0$$

Thus in this example the EC is obtained from (A.6) when $\dot{x}$ and x are identical to zero, i.e., $u_{eq} = -b^{-1}\gamma(t)$. We immediately notice that the disturbance $\gamma(t)$ might be estimated by means of the equivalent control; a way to do it will be given below.

Notice that with the control u being a signum function the right-hand side of (A.6) is not Lipschitz; therefore, we cannot resort to the standard theory of differential equations. To overcome such a complexity, we can use the theory of differential inclusions treated extensively in [15]. Thus, we can obtain a solution of (A.6) in the Filippov sense.

Nevertheless, the effects of real devices, let's say small delays, uncertainties, hysteresis, digital computations, etc., always make it impossible to achieve the identity $s(x) \equiv 0$, and so the trajectories are constrained to some region around the origin, i.e., $\|s(x)\| \leq \Delta$. This is why we can ask for the limit solution of (A.2) when Δ tends to zero. That solution is in fact the solution of (A.2) on the sliding mode and it will be found using the equivalent control method, which will be justified by means of Theorem A.1, given below.

Let $\tilde{u}$ be a control for which we obtain the boundary layer $\|s(x)\| \leq \Delta$. We could say that $\tilde{u}$ is the *real control* which we obtain a real sliding mode with. Thus, the dynamic equations are

$$\dot{x}(t) = f(x,t) + B(x,t)\tilde{u}(t) \tag{A.8}$$

Let us denote by x^* the state vector obtained using the EC method, i.e., the trajectories whose dynamics is governed by (A.5). Let us assume that the distance of any point in the set $S_r = \{x : \|s(x)\| \leq \Delta\}$ to the manifold S is estimated by the inequality

$$d(x,S) \leq P\Delta, \text{ for } P > 0$$

Such a number P always exists if all gradients of functions $s_i(x)$ are linearly independent and are lower bounded in the norm by some positive number. In fact the first condition follows from the assumption that $\det(GB) \neq 0$.

Theorem A.1. *Let us assume that the following four conditions are satisfied:*

(1) There is a solution $x(t)$ of system (A.8) which, on the interval $[0,T]$, fulfills the inequality $\|s(x)\| \leq \Delta$.
(2) For the right-hand part of (A.5), rewritten using x^* as

$$\dot{x}^*(t) = f(x^*,t) - B(x^*,t)[G(x^*)B(x^*,t)]^{-1}[G(x^*)f(x^*,t)] \tag{A.9}$$

a Lipschitz constant exists.
(3) Partial derivatives of the function $B(x,t)[G(x)B(x,t)]^{-1}$ with respect to all arguments exist and are bounded in every bounded domain.
(4) For the right-hand part (A.8) there exist positive numbers M and N such that

$$\|f(x,t) + B(x,t)\tilde{u}\| \leq M + N\|x\| \tag{A.10}$$

Then for any pair of solutions to equations (A.9) and (A.8), with their initial conditions satisfying

$$\|x(0) - x^*(0)\| \leq P\Delta$$

there exists a positive number H such that

$$\|x(t) - x^*(t)\| \leq H\Delta \text{ for all } t \in [0, T]$$

Proof. For (A.8) we will obtain the following derivative on time of $s(x)$:

$$\dot{s}(x) = G(x) f(x, t) + G(x) B(x, t) \tilde{u}(t) \tag{A.11}$$

Since we have assumed that $\det(GB) \neq 0$, from (A.11) we obtain that

$$\tilde{u}(t) = [G(x) B(x, t)]^{-1} \dot{s}(x) - [G(x) B(x, t)]^{-1} G(x) f(x, t) \tag{A.12}$$

The substitution of $\tilde{u}(t)$ into (A.8) yields

$$\dot{x} = f - B[GB]^{-1} Gf + B[GB]^{-1} \dot{s} \tag{A.13}$$

Thus, we have that (A.9) and (A.13) differ from a term depending on $\dot{s}$. By integrating, x^* and x can be written by the following integral equations:

$$x^*(t) = x_0^* + \int_0^t \left\{ f(x^*, \tau) - B(x^*, \tau) [G(x^*) B(x^*, \tau)]^{-1} [G(x^*) f(x^*, \tau)] \right\} d\tau \tag{A.14}$$

$$x(t) = x_0 + \int_0^t \left\{ f(x, \tau) - B(x, \tau) [G(x) B(x, \tau)]^{-1} [G(x) f(x, \tau)] \right\} d\tau$$

$$+ \int_0^t B(x, \tau) [G(x) B(x, \tau)]^{-1} \dot{s}(x) d\tau \tag{A.15}$$

Integrating the last term of (A.15) by parts, and taking into account the hypothesis of the theorem, we can obtain the following estimation of the difference of the two solutions:

$$\|x(t) - x^*(t)\| \leq P\Delta + \int_0^t L \|x(\tau) - x^*(\tau)\| d\tau$$

$$+ \left\| B(x, \tau) [G(x) B(x, \tau)]^{-1} s(x) \right\| \Big|_0^t$$

$$+ \int_0^t \left\| \frac{d}{d\tau} B(x, \tau) [G(x) B(x, \tau)]^{-1} \right\| \|s(x)\| d\tau \tag{A.16}$$

By the assumption (A.10), we have that the norm of $x(t)$ is bounded in an interval $[0, T]$, indeed, since

$$\|x(t)\| \leq \|x(0)\| + MT + \int_0^t N \|x(\tau)\| \, d\tau$$

According to the Bellman–Gronwall lemma (see, e.g., [52]) the following inequality is satisfied:

$$\|x(t)\| \leq (\|x(0)\| + MT) e^{NT}, \text{ for all } t \in [0, T] \qquad (A.17)$$

Thus, by the continuity of f and B, and taking into account hypothesis 3 of the theorem, the inequality (A.16) may be represented as follows:

$$\|x(t) - x^*(t)\| \leq Q\Delta + \int_0^t L \|x(\tau) - x^*(\tau)\| \, d\tau$$

where Q is a positive number. Using the Bellman–Gronwall lemma once again, we obtain the inequality

$$\|x(t) - x^*(t)\| \leq Q\Delta e^{LT}$$

Taking $H = Q e^{LT}$, the theorem is proven.

$\square$

Thus, from the theorem we have that $\lim_{\Delta \to 0} x(t) \to x^*(t)$ in a finite interval. This justifies the equivalent control method.

We have said that the equivalent control method might be used to estimate matched disturbances, as in the example where $u_{eq} = -\gamma$. Next, we will see how to estimate the function u_{eq} by means of a first-order low-pass filter. We will make use of the following lemma.

Lemma A.1. *Let the differential equation be as follows:*

$$\tau \dot{z}(t) + z(t) = h(t) + H(t) \dot{s} \qquad (A.18)$$

where τ is a scalar constant and z, h, and s are m-dimensional function vectors. If the following assumptions are satisfied:

(i) the functions $h(t)$ and $H(t)$ and their first-order derivatives are bounded in magnitude by a certain number M and
(ii) $\|s(t)\| \leq \Delta$, Δ being a constant positive value,

then, for any pair of positive numbers Δt and ε, there exists a number $\delta = \delta(\varepsilon, \Delta t, z(0))$ such that the following inequality is fulfilled:

$$\|z(t) - h(t)\| \leq \varepsilon$$

provided that $0 < \tau \leq \delta$, $\Delta/\tau \leq \delta$ and $t \geq \Delta t$.

A.2 Equivalent Control Method 129

Proof. The solution of (A.18) is as follows:

$$z(t) = e^{-t/\tau} z(0) + \frac{1}{\tau} \int_0^t e^{-(t-\sigma)/\tau} [h(\sigma) + H(\sigma) \dot{s}(\sigma)] d\sigma$$

Integrating by parts we obtain

$$z(t) = e^{-t/\tau} z(0) + h(t) - h(0) e^{-t/\tau}$$

$$- \int_0^t e^{-(t-\sigma)/\tau} \dot{h}(\sigma) d\sigma + H(t) \frac{s}{\tau} - H(0) e^{-t/\tau} \frac{s(0)}{\tau}$$

$$- \frac{1}{\tau} \int_0^t e^{-(t-\sigma)/\tau} \left[\dot{H}(\sigma) + \frac{1}{\tau} H(\tau) \right] s(\sigma) d\sigma$$

Then, by assumptions (i) and (ii), we deduce the following inequality:

$$\|z(t) - h(t)\| \le \|z(0) - h(0)\| e^{-t/\tau} + M\tau + \frac{2M\Delta}{\tau} + M\Delta + \frac{M\Delta}{\tau}$$

Grouping similar terms together yields

$$\|z(t) - h(t)\| \le \|z(0) - h(0)\| e^{-t/\tau} + M(\tau + \Delta) + 3M \frac{\Delta}{\tau} \qquad (A.19)$$

Therefore, it is easy to conclude from (A.19) that for any positive number Δt, the following identity is achieved:

$$\lim_{\substack{\tau \to 0 \\ \Delta/\tau \to 0}} z(t) = h(t) \quad \text{for all } t \ge \Delta t \qquad (A.20)$$

Thus, the lemma is proven.

□

From (A.20), we see that Δ should be much smaller than τ in order to achieve a good estimation of $h(t)$ by means of $z(t)$. Furthermore, (A.19) gives us a more qualitative expression to measure the effect of τ on the estimation. That is, there we can see that if τ is too small, then the term depending on the difference on the initial conditions could be considered negligible, i.e., $z(t)$ reaches rapidly a neighborhood around $h(t)$ of order $O(\tau + \Delta) + O(\frac{\Delta}{\tau})$. In this case, if Δ is not much smaller than τ, then the neighborhood around $h(t)$ will be big. On the other hand if $\Delta \ll \tau$, but τ is not too small, then $z(t)$ would take some time before reaching a small neighborhood around $h(t)$. That is why we can say that an 'ideal' case is when $\Delta \ll \tau \ll 1$.

Thus, the filter designed as

$$\tau \dot{u}_{\text{av}}(t) + u_{\text{av}}(t) = \tilde{u}(t) \qquad (A.21)$$

can be used to estimate u_{eq}. Indeed, from (A.4) and (A.12), (A.21) takes the form

$$\tau u_{\text{av}}(t) + u_{\text{av}}(t) = u_{\text{eq}} + [G(x) B(x,t)]^{-1} \dot{s}(x) \qquad (A.22)$$

Hence, by comparing (A.18) with (A.22), lemma implies that

$$\lim_{\substack{\tau \to 0 \\ \Delta/\tau \to 0}} u_{\text{av}} = u_{\text{eq}} \text{ for } t \in (0, T] \qquad (A.23)$$

provided that u_{eq} and $(GB)^{-1}$ are bounded and have bounded derivatives, which is fulfilled if conditions of Theorem A.1 are fulfilled.

Remark A.1. Let us assume that Δ is known (which in general might not be true). Then, we could select $\tau = \Delta^{1/r}$ $(r > 1)$, implying that $\Delta/\tau = \Delta^{\frac{r-1}{r}}$. Thus, as Δ tends to zero, Δ/τ tends to zero also. Therefore, in that case, (A.23) is still satisfied. For the same qualitative arguments given above, a good estimation of u_{eq} using u_{av} is obtained when $\Delta << \tau << 1$. When r is close to 1 then τ is close to Δ; therefore, r near 1 is not a good selection. On the other hand, for $r >> 1$, τ is close to 1; then in that case r is not a good choice either. By selecting $r = 2$, we obtain, for Δ small enough, that $\Delta << \tau << 1$. Hence, selecting $\tau = \Delta^{1/2}$ and provided that Δ is much smaller than 1, we obtain a good estimation of u_{eq}.

Appendix B

Min–Max Multimodel LQ Control

Abstract This chapter develops a numerical method for the optimal weight adjustment for the min–max LQ problem, where "max" is taken over a finite set of indices (models) and "min" is taken over the set of admissible controls. The control turns out to be a linear combination of the controls that are each one an optimal control when each model is considered individually. Dealing with the control design for some uncertain systems, there exist situations when the model of the system cannot be defined exactly since the more adequate model can depend on several possible scenarios. In this case the control can be designed as a *multimodel control*. To design such control, the min–max approach has been suggested where "max" is taken over all possible models (scenarios) and "min" is taken over all admissible controls. Such *robust optimal control* is shown to be a weighted combination of the controls optimal for each individual model. Hence, the problem is reduced to a finite-dimensional optimization problem since this robust optimal control depends on a weighting vector belonging to the N-dimensional simplex which should be selected providing a minimal value for the original worst LQ functional. Finding an analytical expression for this function as a function of the weights seems to be a very difficult task. In the simplest cases with two (N = 2) and three (N = 3) models, such expression can be easily obtained in a graphic form using a standard PC. However, for more complex situations (N ≥ 4), such cannot be realized. That is why here we give a numerical procedure for the corresponding weight adjustment (optimization).

B.1 Multimodel System

Let us consider a set of linear state models given by

$$\dot{x}^\alpha(t) = A^\alpha(t) x(t) + B^\alpha(t) u(t) + d^\alpha(t), \quad x^\alpha(0) = x_0^\alpha \quad \text{(B.1)}$$

where the index α belongs to a finite set, that is, $\alpha \in \overline{1,N}$ (N is a positive integer), $x^\alpha(t), d^\alpha(t) \in \mathbb{R}^n$, $u(t) \in \mathbb{R}^m$, and $A^\alpha(t)$, $B^\alpha(t)$, and $d^\alpha(t)$ are **known continuous functions** on $t \in [0,T]$. Let us define the performance index as

$$h^\alpha := \frac{1}{2}\left(x^\alpha(t_f), G^\alpha x^\alpha(t_f)\right) + \frac{1}{2}\int_{t=0}^{t_f} \left[(x^\alpha(t), Q^\alpha x^\alpha(t)) + (u(t), Ru(t))\right] dt \tag{B.2}$$

where $Q^\alpha \geq 0$, $G^\alpha \geq 0$, and $R > 0$. The min–max linear quadratic (LQ) control problem is formulated as

$$u^* = \min_{u \in \mathbb{R}^m} \max_{\alpha \in \overline{1,N}} h^\alpha. \tag{B.3}$$

The solution of this problem is as follows.[1] Define the extended system

$$\dot{\mathbf{x}}(t) = \mathbf{A}\mathbf{x}(t) + \mathbf{B}u(\mathbf{x},t) + \mathbf{d}$$

where

$$\mathbf{x} := \begin{bmatrix} x^1 \\ \vdots \\ x^N \end{bmatrix}, \quad \mathbf{A} := \mathrm{diag}\left(A^1(t), \ldots, A^N(t)\right)$$

$$\mathbf{B} := \begin{bmatrix} B^1(t) \\ \vdots \\ B^N(t) \end{bmatrix}, \quad \mathbf{d} := \begin{bmatrix} d^1(t) \\ \vdots \\ d^N(t) \end{bmatrix} \tag{B.4}$$

$$\mathbf{Q} := \mathrm{diag}(Q_1, \ldots, Q_N), \quad \mathbf{G} := \mathrm{diag}(G_1, \ldots, G_N)$$
$$\mathbf{\Lambda} := \mathrm{diag}(\lambda_1 I_{n \times n}, \ldots, \lambda_N I_{n \times n})$$

where $\lambda = (\lambda_1, \ldots, \lambda_N)$ belongs to the simplex $\mathbb{S}^N$ defined as follows:

$$\mathbb{S}^N = \left\{ \lambda \in \mathbb{R}^N : \lambda_i \geq 0, \sum_{i=1}^N \lambda_i = 1 \right\}$$

Then, the *robust optimal control* realizing (B.3) is of the form

$$u = -R^{-1}\mathbf{B}^\intercal \left(\mathbf{P}_\lambda \mathbf{x} + \mathbf{p}_\lambda\right) \tag{B.5}$$

where the matrix $\mathbf{P}_\lambda = \mathbf{P}_\lambda^T \in \mathbb{R}^{nN \times nN}$ is the solution of the parameterized differential matrix Riccati equation:

$$\dot{\mathbf{P}}_\lambda + \mathbf{P}_\lambda \mathbf{A} + \mathbf{A}^T \mathbf{P}_\lambda - \mathbf{P}_\lambda \mathbf{B} R^{-1} \mathbf{B}^T \mathbf{P}_\lambda + \mathbf{\Lambda}\mathbf{Q} = 0; \quad \mathbf{P}_\lambda(T) = \mathbf{\Lambda}\mathbf{G} \tag{B.6}$$

---

[1] For details of how the solution of the problem can be found see Chap. 5.

and the shifting vector $\mathbf{p}_\lambda \in \mathbb{R}^{nN}$ satisfies

$$\dot{\mathbf{p}}_\lambda + \mathbf{A}^\mathsf{T}\mathbf{p}_\lambda - \mathbf{P}_\lambda \mathbf{B} R^{-1}\mathbf{B}^\mathsf{T}\mathbf{p}_\lambda + \mathbf{P}_\lambda \mathbf{d} = \mathbf{0}; \quad \mathbf{p}_\lambda(T) = 0$$

Thus, the solution of (B.3) is reduced to finding the optimal weighting vector λ^* which solves the following finite-dimensional optimization problem:

$$\lambda^* = \arg\min_{\lambda \in \mathbb{S}^N} J(\lambda) \tag{B.7}$$

$$\begin{aligned}
J(\lambda) &:= \max_{\alpha=\overline{1,N}} h^\alpha = \frac{1}{2}\mathbf{x}^T(0)\mathbf{P}_\lambda(0)\mathbf{x}(0) + \mathbf{x}^T(0)\mathbf{p}_\lambda(0) + \\
&+ \frac{1}{2}\max_{\alpha=\overline{1,N}}\left[x^{\alpha T}(t_f)G^\alpha x^\alpha(t_f) + \int_{t=0}^{t_f} x^{\alpha T}(t)Q^\alpha x^\alpha(t)dt\right] - \\
&- \frac{1}{2}\sum_{\alpha=1}^{N}\lambda_i\left[x^{\alpha T}(t_f)G^\alpha x^\alpha(t_f) + \int_{t=0}^{t_f} x^{\alpha T}(t)Q^\alpha x^\alpha(t)dt\right] + \\
&+ \frac{1}{2}\int_{t=0}^{t_f} \mathbf{p}_\lambda^T\left[2\mathbf{d} - \mathbf{B}R^{-1}\mathbf{B}^T\mathbf{p}_\lambda\right]dt
\end{aligned} \tag{B.8}$$

B.2 Numerical Method for the Weight Adjustment

The next step is to develop a numerical method which allows to find the optimal weighting vector λ^* for any finite number N of possible models (or scenarios).

Lemma B.1. *Let λ^* be a minimum point, that is, $J(\lambda^*) \leq J(\lambda)$ for all $\lambda \in \mathbb{S}^N$. Then, for any active index $\alpha \in \overline{1,N}$ such that $1 \geq \lambda_\alpha^* > 0$, the functional $h^\alpha(\lambda^*)$ satisfies the following equality:*

$$h^\alpha(\lambda^*) = J(\lambda^*) \tag{B.9}$$

and for all inactive indices α such that $\lambda_\alpha^* = 0$

$$h^\alpha(\lambda^*) \leq J(\lambda^*) \tag{B.10}$$

Proof. Suppose that for a $j \in \overline{1,N}$ we have $h^j(\lambda^*) > J(\lambda^*)$. Then

$$J(\lambda^*) = \max_{\alpha \in \overline{1,N}} h^\alpha(\lambda^*) \geq h^j(\lambda^*) > J(\lambda^*)$$

which leads to a contradiction. Hence, for all indices α, it follows that $h^\alpha(\lambda^*) \leq J(\lambda^*)$. The result (B.9) for active indices follows directly from the complementary slackness condition established in [53] (see also Chap. 5). □

Corollary B.1. *The optimal performance index $J(\lambda^*)$ can be represented as*

$$J(\lambda^*) = \frac{1}{2}\mathbf{x}^T(0)\mathbf{P}_{\lambda^*}(0)\mathbf{x}(0) + \mathbf{x}^T(0)\mathbf{p}_{\lambda^*}(0) +$$
$$+ \frac{1}{2}\int_{t=0}^{t_f} \mathbf{p}_{\lambda^*}^T(t)\left[2\mathbf{d}(t) - \mathbf{B}(t)R^{-1}\mathbf{B}^T(t)\mathbf{p}_{\lambda^*}(t)\right] dt \quad (B.11)$$

Proof. Adding and subtracting the integral of $u^T(t)Ru(t)$ in (B.8), we get

$$J(\lambda) = \frac{1}{2}\mathbf{x}^T(0)\mathbf{P}_\lambda(0)\mathbf{x}(0) + \mathbf{x}^T(0)\mathbf{p}_\lambda(0) +$$
$$+ \left[J(\lambda) - \sum_{i=1}^{N}\lambda_\alpha h^\alpha(\lambda)\right] + \frac{1}{2}\int_{t=0}^{t_f} \mathbf{p}_\lambda^T\left[2\mathbf{d} - \mathbf{B}R^{-1}\mathbf{B}^T\mathbf{p}_\lambda\right] dt$$

Therefore, taking $\lambda = \lambda^*$, in view of (B.9), and since $\sum_{\alpha=1}^{N}\lambda_\alpha = 1$, we find that $J(\lambda^*) = \sum_{i=1}^{N}\lambda_\alpha^* h^\alpha(\lambda^*)$. Hence the performance index $J(\lambda^*)$ is exactly as it is expressed in (B.11).

$\square$

Corollary B.2. *If the vector λ^* is a minimum point, then for any $\gamma > 0$*

$$\lambda^* = \pi\{\lambda^* + \gamma\mathbf{h}(\lambda^*)\} \quad (B.12)$$

where $\pi\{\cdot\}$ is the projector to the simplex $\mathbb{S}^N$, that is,

$$\|\pi\{x\} - x\| < \|\lambda - x\| \text{ for any } \lambda \in \mathbb{S}^N, \lambda \neq \pi\{x\}$$

and $\mathbf{h}(\lambda) \in \mathbb{R}^N$ is the vector whose ith term is the performance functional h^i, i.e.,

$$\mathbf{h}(\lambda) = \begin{bmatrix} h^1(\lambda) \\ \vdots \\ h^N(\lambda) \end{bmatrix}$$

Proof. Since $\mathbb{S}^N$ is a closed convex set, the following property holds:

for any $x \in \mathbb{R}^n$, $\mu = \pi\{x\} \iff (x - \mu, \lambda - \mu) \leq 0$ for all $\lambda \in \mathbb{S}^N$ (B.13)

Let $\lambda_{i_j}^*$, $j = \overline{1,r}$ be the components of λ^* different from zero and $\lambda_{i_k}^*$ $k = \overline{r+1,N}$ be the components of λ^* equal to zero. Thus, taking into account Lemma B.1 and since $\lambda_{i_k} - \lambda_{i_k}^* \geq 0$ ($\lambda_{i_k}^* = 0$), we obtain

B.2 Numerical Method for the Weight Adjustment

$$\left(\lambda^* + \gamma \mathbf{h}\left(\lambda^*\right) - \lambda^*, \lambda - \lambda^*\right) =$$

$$= \gamma \left[J\left(\lambda^*\right) \sum_{j=1}^{r} \left(\lambda_{i_j} - \lambda_{i_j}^*\right) + \sum_{k=r+1}^{N} h^{i_k}\left(\lambda^*\right)\left(\lambda_{i_k} - \lambda_{i_k}^*\right) \right] \leq$$

$$\leq \gamma J\left(\lambda^*\right) \left[\sum_{j=1}^{r} \left(\lambda_{i_j} - \lambda_{i_j}^*\right) + \sum_{k=r+1}^{N} \left(\lambda_{i_k} - \lambda_{i_k}^*\right) \right] = \gamma J\left(\lambda^*\right) \sum_{j=1}^{N} \left(\lambda_{i_j} - \lambda_{i_j}^*\right) = 0$$

(B.14)

for all $\lambda \in \mathbb{S}^N$. Thus, by (B.13), (B.14) implies $\lambda^* = \pi\{\lambda^* + \gamma \mathbf{h}(\lambda^*)\}$.

□

In [53] (see Chap. 5) it was shown that the control $u(\mathbf{x},t)$ designed as in (B.3) is the combination (where the weights are the components λ_α) of the controls optimal for each individual model. Hence, it seems to be clear that a bigger weight λ_α of the control, optimizing the α-model, implies a better (smaller) performance index $h^\alpha(\lambda)$. This fact may be expressed in the following manner: if $\lambda'_\alpha \neq \lambda''_\alpha$

$$\left(\lambda'_\alpha - \lambda''_\alpha\right)\left[h^\alpha\left(\lambda'\right) - h^\alpha\left(\lambda''\right)\right] < 0 \tag{B.15}$$

for any $\lambda' \neq \lambda'' \in \mathbb{S}^N$. Adding (B.15) on $\alpha \in \overline{1,N}$ leads to the following condition which we will accept as an assumption.

A6.1 For any $\lambda' \neq \lambda'' \in \mathbb{S}^N$ the following inequality holds

$$\left(\lambda' - \lambda'', \mathbf{h}\left(\lambda'\right) - \mathbf{h}\left(\lambda''\right)\right) < 0 \tag{B.16}$$

and the identity in (B.16) is possible only if $\lambda' = \lambda''$.

Proposition B.1. *Under A6.1, the functional $J(\lambda)$ has a unique minimum point λ^*.*

Proof. We will show that if $\tilde{\lambda}$ differs from λ^*, then $\tilde{\lambda}$ does not satisfy the identity (B.12).

Let us assume that $\tilde{\lambda} \neq \lambda^*$. Then (B.14) implies

$$\left(\tilde{\lambda} + \gamma \mathbf{h}\left(\tilde{\lambda}\right) - \tilde{\lambda}, \lambda^* - \tilde{\lambda}\right) \geq$$

$$\geq \gamma \left[\left(\mathbf{h}\left(\tilde{\lambda}\right), \lambda^* - \tilde{\lambda}\right) + \left(\mathbf{h}\left(\lambda^*\right), \tilde{\lambda} - \lambda^*\right)\right] = \gamma \left(\mathbf{h}\left(\tilde{\lambda}\right) - \mathbf{h}\left(\lambda^*\right), \lambda^* - \tilde{\lambda}\right)$$

(B.17)

On the other hand, A6.1 yields the following:

$$\gamma \left(\mathbf{h}\left(\tilde{\lambda}\right) - \mathbf{h}\left(\lambda^*\right), \lambda^* - \tilde{\lambda}\right) = -\gamma \left(\tilde{\lambda} - \lambda^*, \mathbf{h}\left(\tilde{\lambda}\right) - \mathbf{h}\left(\lambda^*\right)\right) > 0 \tag{B.18}$$

Both (B.17) and (B.18) imply

$$\left(\tilde{\lambda} + \gamma \mathbf{h}\left(\tilde{\lambda}\right) - \tilde{\lambda}, \lambda^* - \tilde{\lambda}\right) > 0 \tag{B.19}$$

Nevertheless, (B.19) means that $\tilde{\lambda} \neq \pi\left\{\tilde{\lambda} + \gamma \mathbf{h}\left(\tilde{\lambda}\right)\right\}$ [see (B.13)]. Therefore, by Corollary B.2, we can deduce that $\tilde{\lambda}$ is not a minimum point. □

Now, we are ready to present a numerical method for the adjustment of the weight vector λ.

B.2.1 Numerical Method

Define the sequence of vectors $\left\{\lambda^k\right\}$ as

$$\lambda^{k+1} = \pi\left\{\lambda^k + \frac{\gamma^k}{J\left(\lambda^k\right) + \varepsilon} \mathbf{h}\left(\lambda^k\right)\right\}, \lambda^0 \in \mathbb{S}^N, k = 0, 1, 2, \ldots$$
$$\mathbf{h}\left(\lambda^k\right) = \left[h^1\left(\lambda^k\right) \cdots h^N\left(\lambda^k\right)\right]^T \quad (B.20)$$
$$J\left(\lambda^k\right) := \max_{\alpha \in \overline{1,N}} h^\alpha\left(\lambda^k\right)$$

where ε is an arbitrary strictly positive (small enough) constant.

Theorem B.1. *Let λ^* be the minimum point for $J(\lambda)$. If*

(1) the sequence $\left\{\lambda^k\right\}$ is generated by (B.20)
(2) A6.1 holds
(3) there exists a constant L such that for all $\alpha \in \overline{1,N}$ and for any $\mu, \lambda \in \mathbb{S}^N$

$$|h^\alpha(\mu) - h^\alpha(\lambda)| \leq J(\lambda) L |\mu - \lambda|$$

(4) the gain sequence $\{\gamma^k\}$ satisfies

$$\gamma^k > 0, \ \sum_{k=0}^\infty \gamma^k = \infty, \ \sum_{k=0}^\infty \left(\gamma^k\right)^2 < \infty$$

then

$$\lim_{k \to \infty} \lambda^k = \lambda^*. \quad (B.21)$$

Proof. For $v^k := \lambda^k - \lambda^*$, in view of (B.12) and the property of projection $\|\pi\{x\} - \pi\{y\}\| \leq \|x - y\|$ for all $x, y \in \mathbb{R}^N$, we obtain

B.2 Numerical Method for the Weight Adjustment

$$\left\| v^{k+1} \right\|^2 = \left\| \pi \left\{ \lambda^k + \frac{\gamma^k}{J(\lambda^k) + \varepsilon} \mathbf{h}(\lambda^k) \right\} - \lambda^* \right\|^2 =$$

$$= \left\| \pi \left\{ \lambda^k + \frac{\gamma^k}{J(\lambda^k) + \varepsilon} \mathbf{h}(\lambda^k) \right\} - \pi \left\{ \lambda^* + \frac{\gamma^k}{J(\lambda^k) + \varepsilon} \mathbf{h}(\lambda^*) \right\} \right\|^2 \le$$

$$\le \left\| v^k + \frac{\gamma^k}{J(\lambda^k) + \varepsilon} \left[\mathbf{h}(\lambda^k) - \mathbf{h}(\lambda^*) \right] \right\|^2 =$$

$$= \left\| v^k \right\|^2 + \frac{(\gamma^k)^2}{J^2(\lambda^k) + 2J(\lambda^k)\varepsilon + \varepsilon^2} \left\| \mathbf{h}(\lambda^k) - \mathbf{h}(\lambda^*) \right\|^2 +$$

$$+ 2 \frac{\gamma^k}{J(\lambda^k) + \varepsilon} \left(v^k, \mathbf{h}(\lambda^k) - \mathbf{h}(\lambda^*) \right) \le$$

$$\le \left\| v^k \right\|^2 \left(1 + (\gamma^k)^2 L^2 \right) + 2 \frac{\gamma^k}{J(\lambda^k) + \varepsilon} \left(v^k, \mathbf{h}(\lambda^k) - \mathbf{h}(\lambda^*) \right) \le$$

$$\le \left\| v^k \right\|^2 \left(1 + (\gamma^k)^2 L^2 \right).$$
(B.22)

In the last inequality of (B.22) we have used A6.1. Define the new variable w^k by

$$w^k := \left\| v^k \right\|^2 \prod_{s=k}^{\infty} \left[1 + (\gamma^s)^2 L^2 \right].$$

With the previous definition of w^k, (B.22) implies

$$w^{k+1} := \left\| v^{k+1} \right\|^2 \prod_{s=k+1}^{\infty} \left[1 + (\gamma^s)^2 L^2 \right] \le$$

$$\le \left\| v^k \right\|^2 \left(1 + (\gamma^k)^2 L^2 \right) \prod_{s=k+1}^{\infty} \left[1 + (\gamma^s)^2 L^2 \right] = w^k$$

which means (by Weierstrass theorem) that the sequence $\{w^k\}$ converges and, hence, implies the existence of the limit

$$w := \lim_{k \to \infty} w^k = \lim_{k \to \infty} \left\| v^k \right\|^2$$

Nevertheless, from (B.22), we also have the inequality

$$2 \frac{\gamma^k}{J(\lambda^k) + \varepsilon} \left| \left(v^k, \mathbf{h}(\lambda^k) - \mathbf{h}(\lambda^*) \right) \right| \le \left\| v^k \right\|^2 \left(1 + (\gamma^k)^2 L^2 \right) - \left\| v^{k+1} \right\|^2 =$$

$$\frac{w^k - w^{k+1}}{\prod_{s=k+1}^{\infty} \left[1 + (\gamma^s)^2 L^2 \right]} \le w^k - w^{k+1}$$

Summation of it by k from 0 up to ∞ yields

$$2\sum_{k=0}^{\infty}\gamma^{k}\frac{\left|\left(v^{k},\mathbf{h}\left(\lambda^{k}\right)-\mathbf{h}\left(\lambda^{*}\right)\right)\right|}{J\left(\lambda^{k}\right)+\varepsilon}\leq w^{0}-w<\infty$$

In view of the property $\sum_{k=0}^{\infty}\gamma^{k}=\infty$, it follows that there exists a subsequence k_t $(t=1,2,\ldots)$ such that

$$\frac{\left|\left(v^{k_t},\mathbf{h}\left(\lambda^{k_t}\right)-\mathbf{h}\left(\lambda^{*}\right)\right)\right|}{J\left(\lambda^{k_t}\right)+\varepsilon}\underset{t\to\infty}{\longrightarrow}0 \tag{B.23}$$

Since $J\left(\lambda^{k_t}\right)$ is bounded, then, by (B.16), the limit in (B.23) implies that $\lambda^{k_t}\underset{t\to\infty}{\longrightarrow}\lambda^{*}$, or, equivalently,

$$\lim_{t\to\infty}w^{k_t}=\lim_{t\to\infty}\left\|v^{k_t}\right\|^{2}=0$$

Since $\{w^k\}$ converges to w, all its subsequences converge to the same limit, which in turn implies that $w = 0$. Theorem is proven. □

B.3 Example

The following example illustrates the proposed numerical method (B.20) in the case $N = 3$ where the parameters of possible models are as follows:

$$A_1 = \begin{bmatrix} -2 & 0.5 & 1 \\ 0.5 & 1.2 & -2 \\ 1 & 2 & -1.5 \end{bmatrix}, A_2 = \begin{bmatrix} -0.3 & 1.5 & -0.15 \\ -1 & 0.12 & 2 \\ 1 & 2 & -3 \end{bmatrix}, A_3 = \begin{bmatrix} 0.4 & -1 & 0.3 \\ 0.5 & -0.4 & 0.3 \\ 0.5 & 0.6 & -1 \end{bmatrix}$$

$$B_1 = \begin{bmatrix} 0.5 \\ 1 \\ 1 \end{bmatrix}, B_2 = \begin{bmatrix} 1.5 \\ -2 \\ 1 \end{bmatrix}, B_3 = \begin{bmatrix} 0.5 \\ 0.2 \\ 1 \end{bmatrix}$$

$$d^1 = \begin{bmatrix} 0.1 \\ 0.05 \\ 0.01 \end{bmatrix}, d^2 = \begin{bmatrix} 0.1\sin(t) \\ 0.2\sin(t/2) \\ 0.1 \end{bmatrix}, d^3 = \begin{bmatrix} 0.1 \\ 0.05\cos(t) \\ 0.1 \end{bmatrix}$$

For simulation purposes, we select matrices $Q^\alpha = G^\alpha = I$, $R = 1$. Using the gain-step sequence $\{\gamma^k\}$ given in (B.20) with $\gamma^k = \dfrac{1}{k+1}$ $(k = 0, 1, 2, \ldots)$, we obtained the results presented in Table B.1. There, the values of the vector λ^k and the performance index $h^\alpha\left(\lambda^k\right)$ are shown for each iteration k.

From Table B.1, one can see that the weights practically converge after 10 iterations. Since all indices are active ($\lambda_\alpha^* > 0$), all performance functionals $h^\alpha\left(\lambda^k\right)$ practically turn out to be equal after 40 iterations. Thus, we have $\lambda^* \approx (0.072035, 0.296663, 0.631301)$. The control law $u = u\left(\lambda^*\right)$ is depicted in Fig. B.1. Figures B.2–B.4 show the trajectories of x^α for $\alpha = 1, 2, 3$.

Table B.1. Values of λ^k and $h^\alpha\left(\lambda^k\right)$

k	λ_1	λ_2	λ_3	h^1	h^2	h^3	J
1	0.5	0.4	0.1	116.060	109.897	977.037	977.037
2	0.208365	0.102057	0.689576	236.303	1116.18	310.533	1116.18
3	0.065899	0.353738	0.580362	489.798	349.164	503.783	503.783
4	0.093832	0.288619	0.617548	360.415	462.742	460.030	462.742
5	0.057465	0.307535	0.634999	555.476	432.953	451.644	555.476
6	0.084632	0.290587	0.624780	393.174	461.878	455.737	461.878
7	0.068843	0.299589	0.631567	471.404	447.301	453.104	471.404
8	0.073125	0.296569	0.630305	446.649	452.593	453.386	453.386
9	0.071960	0.297042	0.630997	453.083	451.886	452.979	453.083
10	0.072066	0.296855	0.631077	452.488	452.251	452.875	452.875
⋮	⋮	⋮	⋮	⋮	⋮	⋮	⋮
31	0.072036	0.296664	0.631299	452.660	452.656	452.665	452.665
32	0.072036	0.296664	0.631299	452.661	452.657	452.664	452.664
33	0.072036	0.296663	0.631300	452.666	452.658	452.664	452.665
34	0.072036	0.296663	0.631300	452.661	452.658	452.664	452.664
35	0.072036	0.296663	0.631300	452.660	452.658	452.664	452.664
36	0.072035	0.296663	0.631300	452.660	452.660	452.663	452.663
37	0.072035	0.296663	0.631300	452.661	452.659	452.663	452.663
38	0.072035	0.296663	0.631300	452.660	452.660	452.663	452.663
39	0.072035	0.296663	0.631301	452.662	452.659	452.663	452.663
40	0.072035	0.296663	0.631301	452.662	452.659	452.663	452.663

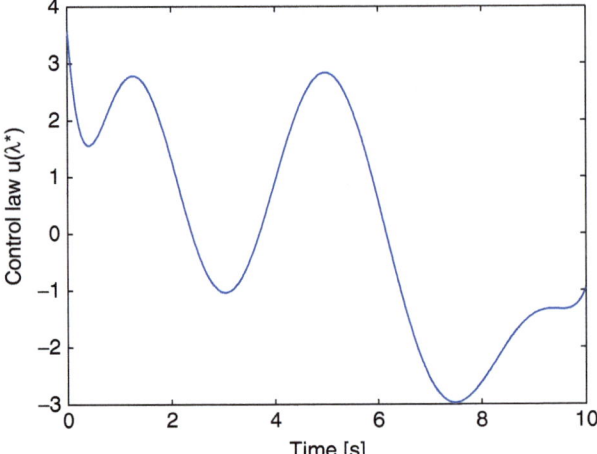

Fig. B.1. Control law u for λ^*.

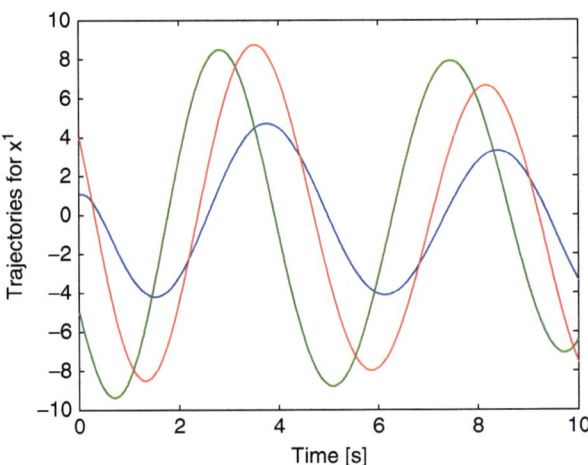

Fig. B.2. Trajectories of the state corresponding to $\alpha = 1$.

B.3 Example 141

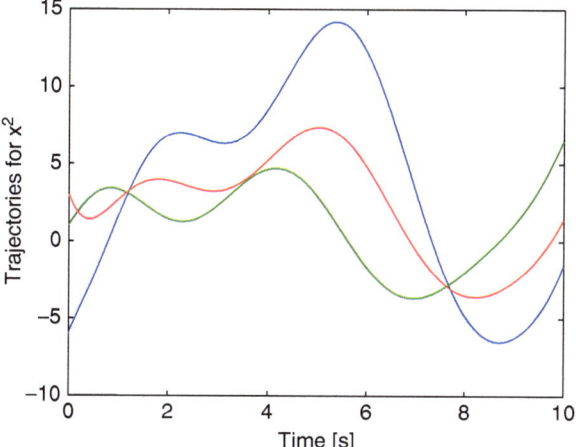

Fig. B.3. Trajectories of the state corresponding to $\alpha = 2$.

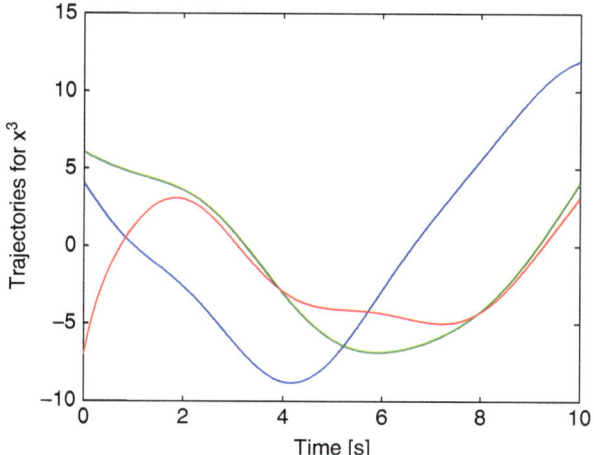

Fig. B.4. Trajectories of the state corresponding to $\alpha = 3$.

Notations

ODE	-	Ordinary differential equation
SM	-	Sliding mode
ISM	-	Integral sliding mode
OISM	-	Output integral sliding mode
$\mathbb{R}$	-	The field of real numbers
$\mathbb{R}^n$	-	The vector space of dimension equal to n
$\mathbb{C}$	-	The field of complex numbers
$\mathbb{C}^-$	-	The set of complex numbers with negative real part
(x,y)	$= x^T y$	
$\|x\|$	$= \sqrt{x^T x}$ (the Euclidean norm)	
$\phi(t)$	-	Represents a matched uncertainty, i.e., $\phi(t) = B\gamma(t)$ for some function $\gamma(t)$
$\mathrm{diag}(X_1, X_2, \ldots, X_r)$	-	A block diagonal matrix with the matrices X_1, X_2, to X_r in the main diagonal blocks and zeros elsewhere
$B^\top$	-	The transpose of B matrix
$B^\perp$	-	A matrix whose transposed rows form a basis of the orthogonal space of $\mathrm{Im}\, B$ ($B^\perp B = 0$)
B^+	-	The Moore–Penrose pseudoinverse of B
I_m	-	Identity matrix of dimension m by m
P_λ	-	Parameterized Riccati matrix
$s(x)$	-	Sliding variable
u_{eq}	-	Equivalent control.
$\lambda_{\max}(A)$	-	The greatest eigenvalue of the square A matrix
$\lambda_{\min}(A)$	-	The lowest eigenvalue of the square A matrix

References

1. M. Athans, Editorial on the lqg problem. IEEE Trans. Autom. Control **16**(6), 528 (1971)
2. J. Ackermann, *Robuste Regelung* (Springer, New York, 2011)
3. M. Morari, E. Zafiriou, *Robust Process Control Theory* (Prentice Hall, Englewood Cliffs, 1989)
4. M.G. Safonov, *Stability and Robustness of Multivariable Feedback Systems* (MIT Press, Cambridge, 1980)
5. K. Zhou, J. Doyle, *Essentials of Robust Control* (Prentice Hall, New Jersey, 1999)
6. D. McFarlane, K. Glover, *Robust Controller Design Using Normalized Coprime Factor Plant Descriptions*. Lectures Notes in Control and Information Sciences (Springer, New York, 1989)
7. L. Ray, R. Stengel, Stochastic robustness of linear-time-invariant control systems. IEEE Trans. Automat. Control **36**(1), 82–87 (1991)
8. S. Bhattacharyya, H. Chapellat, L. Keel, *Robust Control: The Parametric Approach* (Prentice Hall, New Jersey, 1995)
9. V. Boltynaski, A. Poznyak, *The Robust Maximum Principle: Theory and Applications* (Birkhauser, New York, 2012)
10. V. Utkin, First stage of vss: people and events, in *Variable Structure Systems: Towards the 21st Century*, ed. by X. Yu, J.-X. Xu. Lectures Notes in Control and Information Sciences, vol. 274 (Springer, London, 2002), pp. 1–32
11. V. Utkin, *Sliding Modes in Control and Optimization* (Springer, Berlin, 1992)
12. C. Edwards, S.K. Spurgeon, *Sliding Mode Control: Theory and Applications* (CRC Press, New York, 1998)
13. V. Utkin, J. Guldner, J. Shi, *Sliding Mode Control in Electromechanical Systems* (Taylor & Francis Inc., Philadelphia, 1999)
14. Y. Shtessel, C. Edwards, L. Fridman, A. Levant, *Sliding Mode Control and Observation* (Birkhäuser, Boston, 2013)
15. A. Filippov, *Differential Equations with Discontinuous Right-hand Sides* (Kluwer, Dordrecht, 1988)
16. B. Drazenovic, The invariance conditions in variable structure systems. Automatica **5**, 287–295 (1969)
17. A. Lukyanov, V. Utkin, Methods of reducing equations for dynamic systems to a regular form. Autom. Rem. Control **42**(4), 413–420 (1981)

18. X.-G. Yan, S. Spurgeon, C. Edwards, Dynamic sliding mode control for a class of systems with mismatched uncertainty. Eur. J. Control **11**, 1–10 (2005)
19. H. Choi, An lmi-based switching surface design method for a class of mismatched uncertain systems. IEEE Trans. Autom. Control **48**(9), 1634–1638 (2003)
20. A. Estrada, L. Fridman, Quasi-continuous hosm control for systems with unmatched perturbations. Automatica **11**, 1916–1919 (2010)
21. A. Estrada, L. Fridman, Integral hosm semiglobal controller for finite-time exact compensation of unmatched perturbations. IEEE Trans. Autom. Control **55**(11), 2644–2649 (2010)
22. A. Poznyak, Y. Shtessel, C. Gallegos, Mini-max sliding mode control for multi-model linear time varying systems. IEEE Trans. Autom. Control **48**(12), 2141–2150 (2003)
23. G. Herrmann, S. Spurgeon, C. Edwards, A model-based sliding mode control methodology applied to the hda plant. J. Process Control **13**, 129–138 (2003)
24. F. Castaños, L. Fridman, Dynamic switching surfaces for output sliding mode control: an h_∞ approach. Automatica **47**(7), 1957–1961 (2011)
25. A. Ferreira, F. Bejarano, L. Fridman, Unmatched uncertainties compensation based on high-order sliding mode observation. Int. J. Robust Nonlinear Control **23**(7), 754–764 (2013)
26. G.P. Matthews, R.A. DeCarlo, Decentralized tracking for a class of interconnected nonlinear systems using variable structure control. Automatica **24**, 187–193 (1988)
27. V.I. Utkin, J. Shi, Integral sliding mode in systems operating under uncertainty conditions, in *Proceedings of the 35th IEEE-CDC*, Kobe, Japan, 1996
28. F. Castaños, L. Fridman, Analysis and design of integral sliding manifolds for systems with unmatched perturbations. IEEE Trans. Autom. Control **55**(5), 853–858 (2006)
29. M. Rubagotti, A. Estrada, F. Castaños, A. Ferrara, L. Fridman, Integral sliding mode control for nonlinear systems with matched and unmatched perturbations. IEEE Trans. Autom. Control **56**(11), 2699–2704 (2011)
30. J.-X. Xu, W. Cao, Nonlinear integral-type sliding surface for both matched and unmatched uncertain systems. Proc. Am. Control Conf. **6**, 4369–4374 (2001)
31. J. Xu, Y. Pan, T. Lee, L. Fridman, On nonlinear h-infinity sliding mode control for a class of nonlinear cascade systems. Int. J. Syst. Sci. **36**(15), 983–992 (2005)
32. F. Castaños, J. Xu, L. Fridman, Integral sliding modes for systems with matched and unmatched uncertainties. *Advances in Variable Structure and Sliding Mode Control*. Lecture Notes in Control and Information Sciences, vol. 334 (Springer, Berlin, 2006), pp. 227–246
33. F. Bejarano, L. Fridman, A. Poznyak, Output integral sliding mode control based on algebraic hierarchical observer. Int. J. Control **80**(3), 443–453 (2007)
34. F.J. Bejarano, L.M. Fridman, A.S. Poznyak, Output integral sliding mode for min–max optimization of multi-plant linear uncertain systems. IEEE Trans. Autom. Control **54**(11), 2611–2620 (2009)
35. B. Anderson, J. Moore, *Optimal Control: Linear Quadratic Methods*. Information and Systems Science (Prentice Hall, London, 1990)
36. A. Isidori, *Nonlinear Control Systems*, 3rd edn. Communications and Control Engineering (Springer, New York, 1995)
37. L. Fridman, F. Castaños, N. M'Sirdi, N. Khraef, Decomposition and robustness properties of integral sliding mode controllers, in *8th International Workshop on Variable Structure Systems*, Spain, Sept 2004, paper no. f-30

38. A. Poznyak, L. Fridman, F. Bejarano, Mini-max integral sliding mode control for multimodel linear uncertain systems. IEEE Trans. Autom. Control **49**(1), 97–102 (2004)
39. L. Fridman, A. Poznyak, F. Bejarano, Decomposition of the min–max multimodel problem via integral sliding mode. Int. J. Robust Nonlinear Control **15**(13), 559–574 (2005)
40. J.-X. Xu, Y. Pan, T. Lee, Analysis and design of integral sliding mode control based on Lyapunov's direct method. in *Proceedings of the American Control Conference*, Denver, Colorado, June 2003, pp. 192–196
41. C. Edwards, S. Spurgeon, Sliding mode stabilization of uncertain systems using only output information. Int. J. Control **62**(5), 1129–1144 (1995)
42. S. Bag, S. Spurgeon, C. Edwards, Output feedback sliding mode design for linear uncertain systems. IEEE Proc. Control Theor. Appl. **144**, 209–216 (1997)
43. C. Edwards, S. Spurgeon, R.G. Hebden, On development and applications of sliding mode observers, in *Variable Structure Systems: Towards XXIst Century*, ed. by J. Xu, Y. Xu. Lecture Notes in Control and Information Science (Springer, Berlin, 2002), pp. 253–282
44. H. Sira-Ramirez, M. Fliess, On the output feedback control of a synchronous generator, in *IEEE Conference on Decision and Control*, Bahamas, Dec 2004
45. H. Hashimoto, V. Utkin, J. Xu, H. Suzuki, F. Harashima, Vss observer for linear time varying system, in *Proceeding of IECON'90*, Pacific Grove CA, 1990, pp. 34–39
46. J. Barbot, M. Djemai, T. Boukhobza, Implicit triangular observer form dedicated to a sliding mode observer for systems with unknown inputs. Asian J.Control **5**, 513–527 (2003)
47. T. Floquet, J.P. Barbot, A sliding mode approach of unknown input observers for linear systems. in *43th IEEE Conference on Decision and Control*, 2004, pp. 1724–1729
48. C. Chen, *Linear Systems: Theory and Design* (Oxford University Press, New York, 1999)
49. A. Albert, *Regression and Moore-Penrose Pseudoinverse*. Mathematics in Science and Engineering, vol. 94 (Academic, New York, 1972)
50. V. Boltyansky, A. Poznyak, Robust maximum principle in minimax control. Int. J. Control **72**(4), 305–314 (1999)
51. V.G. Boltyanski, A.S. Poznyak, Robust maximum principle for minimax Bolza problem with terminal set, in *Proceedings of IFAC-99, F (2d-06)*, 1999, pp. 263–268
52. A.S. Poznyak, *Advanced Mathematical Tools for Automatic Control Engineers. Vol. 1: Deterministic Technique* (Elsevier, New York, 2008)
53. A. Poznyak, T. Duncan, B. Pasik-Duncan, V. Boltyansky, Robust maximum principle for minimax linear quadratic problem. Int. J. Control **75**(15), 1170–1177 (2002)
54. D.D. Siljak, *Decentralized Control of Complex Systems* (Academic, New York, 1991)
55. A. Linnemann, *Decentralized Control of Dynamically Interconnected Systems* (Universität Bremen, Bremen, 1983)
56. D. Stewart, A plataform with six degrees of freedom. Proc. Inst. Mech. Eng. **180**(15), 371–386 (1965/1966)
57. L. Fraguela, L. Fridman, V. Alexandrov, Output integral sliding mode control to stabilize position a stewart platform. J. Franklin Inst. **349**(4), 1526–1542 (2011)

58. K. Lee, D. Shah, Dynamic analysis of a three degree of freedom in parallel actuated manipulator. IEEE J. Robotics Autom. **4**(3), 361–367 (1988)
59. V. Alexandrov, V. Boltiansky, S. Lemak, N. Parusnikov, V. Tijomirov, *Optimal Control of Motion* (FISMATLIT, Russian, 2005)
60. M. Jungers, A. Franco, E. Pieri, H. Abou-Kandil, Nash strategy applied to active magnetic bearing control, in *Proceedings of IFAC World Congress*, vol. 16, Praha, Czech Republic, 2005
61. A. Ferreira de Loza, M. Jiménez-Lizarraga, L. Fridman, Robust output nash strategies based on sliding mode observation in a two-player differential game. J. Franklin Inst. **349**(4), 1416–1429 (2011)
62. T. Başar, G. Olsder, *Dynamic Noncooperative Game Theory* (SIAM, New York, 2002)
63. H. Abou-Kandil, G. Freiling, V. Ionescu, G. Jank, *Matrix Riccati Equations in Control and System Theory* (Birkhauser, New York, 2003)
64. J. Engwerda, *LQ Dynamic Optimization and Differential Games* (Wiley, West Sussex, 2005)
65. A. Weeren, J. Schumacher, J. Engwerda, Asymptotic analysis of linear feedback nash equilibria in nonzero-sum linear-quadratic differential games. J. Optim. Theor. Appl. **101**, 693–723 (1999)
66. T. Li, Z. Gajic, *New Trends in Dynamic Games and Applications* (Birkhuser, Boston, 1995)
67. J. Ackermann, D. Kaesbaver, W. Sienel, R. Steinhauser, *Robust Control Systems with Uncertain Physical Parameters* (Springer, New York, 1994)
68. R. Murray-Smith, T.A. Johansen, *Multiple Model Approaches to Modeling and Control* (Taylor & Francis, London, 1997)
69. M. Ksouri-Lahmari, A.E. Kamel, M. Benrejeb, P. Borne, Multimodel, multicontrol decision making in system automation, in *IEEE-SMC'97*, Orlando, Florida, USA, Oct 1997
70. J.-F. Magni, Y. Gorrec, C. Chiappa, An observer based multimodel control design approach. Int. J. Syst. Sci. **30**(1), 61–68 (1999)
71. V. Demyanov, V. Malozemov, *Introduction to Minimax* (Dover, New York, 1990)
72. W. Schmitendorf, A sufficient condition for minmax control of systems with uncertainty in the state equations. IEEE Trans. Autom. Control **21**(4), 512–515 (1976)

Index

admissible control, 47
adjoint variables, 48

equivalent control, 14, 17, 33, 81
equivalent output injection, 24, 26, 83, 84, 99

feasible control, 46

Hamiltonian function, 49
hierarchical observer, 22
    integral sliding mode, 27, 32, 36
    sliding mode, 81, 85
HISM, see hierarchical integral sliding mode

integral sliding function, 62
integral sliding mode, 15, 25

low pass filter, 28, 36
LQ index, 61, 63, 68, 74

matched disturbance, 32
matching condition, 60
Mayer problem, 48
min–max LQ control, 63
min-max Bolza problem, 47
min-max control, 45, 86
minimum point, 134–136
multimodel, 68

observer
    Luenberger, 82
OISM, see output integral sliding mode
optimal control, 63, 132
output injection, 23, 35, 99
output integral sliding mode, 79

Riccati equation
    algebraic matrix, 16
    differential matrix, 36
    parameterized differential matrix, 54, 63, 69–71, 86, 132
robust optimal control, 79

shifting vector, 54, 63, 72, 86
simplex, 132
sliding function, 61
system matrix, 34

the cost function, 47

unmatched disturbance, 17

weighting vector, 86, 133
worst (highest) cost, 47

zeros, 34

If you have any concerns about our products,
you can contact us on
ProductSafety@springernature.com

In case Publisher is established outside the EU,
the EU authorized representative is:
**Springer Nature Customer Service Center GmbH
Europaplatz 3, 69115 Heidelberg, Germany**

Printed by Libri Plureos GmbH
in Hamburg, Germany